# THIN AIR

## How Wireless Technology
## Supports Lean Initiatives

# THIN AIR

## How Wireless Technology
## Supports Lean Initiatives

Dann Anthony Maurno

Louis Sirico

CRC Press
Taylor & Francis Group
Boca Raton   London   New York

CRC Press is an imprint of the
Taylor & Francis Group, an **informa** business

A PRODUCTIVITY PRESS BOOK

Productivity Press
Taylor & Francis Group
270 Madison Avenue
New York, NY 10016

© 2010 by Taylor and Francis Group, LLC
Productivity Press is an imprint of Taylor & Francis Group, an Informa business

No claim to original U.S. Government works

Printed in the United States of America on acid-free paper
10 9 8 7 6 5 4 3 2 1

International Standard Book Number: 978-1-4398-0439-1 (Hardback)

| Library of Congress Cataloging-in-Publication Data |
| --- |

Maurno, Dann Anthony.
   Thin air : how wireless technology supports lean initiatives / Dann Anthony Maurno, Louis Sirico.
      p. cm.
   Includes bibliographical references and index.
   ISBN 978-1-4398-0439-1 (hbk. : alk. paper)
   1. Radio frequency identification systems--Industrial applications. 2. Wireless communication systems--Cost effectiveness. 3. Lean manufacturing. 4. Business logistics. I. Sirico, Louis, 1967- II. Title.

TK6570.I34M38 2010
658.5--dc22
                                         2010008210

**Visit the Taylor & Francis Web site at**
**http://www.taylorandfrancis.com**

**and the Productivity Press Web site at**
**http://www.productivitypress.com**

# Contents

# Acknowledgments

It has been a joy and a privilege to interview dozens of industry wizards in researching *Thin Air*. Among them they comprised a star panel of solution providers or results providers (as one of them prefers to be called), Lean consultants, technology users, academics, analysts, and futurists. These include Zander Livingston of American Apparel; Toby Rush of Rush Tracking Systems, who is both a Lean and Wireless enthusiast; Eric Fleming of Rush Tracking Systems, who was instrumental in writing the Lean Wireless ROI calculator; Thomas Pavela of Omni-ID; Carl Brown of Simply RFID; Tuomo Rutanen of Ekahau; Jeff Jacobsen of Kopin; John Wass of WaveMark, Inc., who led us to Six Sigma Black Belt Lynda Wilson of Mercy Heart Hospital; Raj Saksena and David Orain of Omnitrol; Kevin Ashton, a co-founder and former executive director of the Auto-ID Center at MIT; Ravi Pappu, co-founder of ThingMagic; Jamie Flinchbaugh, co-founder of the Lean Learning Center; Chris Schaefer and Chris Warner of Motorola; Kevin Prouty, formerly of Motorola, now with Infor Global Solutions; Dr. Can Saygin of the University of Texas at San Antonio; L. Allen Bennett of Entigral Systems; Michael J. Liard of ABI Research; David Phillips of Wherenet, Inc.; Ralph Rio of ARC Advisory Group, a true industry sage; Marc-Anthony Signorino of the National Association of Manufacturers; Ganesh Wadawadigi of SAP Labs; author and futurist Jamais Cascio; Chris Parker of Octave Technology; Carlos Arteaga; Professor William Gribbons of Bentley College; J. P. McCormick of Activeworlds; Tim Casey of Intel, Inc.; David Laux and George Dolbier of IBM Games and Interactive Entertainment; Jason Howe at Awarepoint; and Joe Leone and John Kuester of RFID Global Solution among many, many others.

Special thanks as well to the long list of connectors and professionals who helped us to assemble the star panel, including Lauren Berg of Edelman, Chicago; Kerry Farrell, also of Edelman; David Fretwell of PAN Communications; Kim Novino and Roger Bridgeman of Bridgeman Communications; Kimberly Kennedy of Perkett PR; Daniella Seghieri of TechInsights; Emily Murphy of Shift Communications; Lynda Kaye of Kaye Public Relations; and Ross Perich of Trainer Communications. Yes, as so many of you told us, you were just doing your jobs, but you do them so well! Your clients and we appreciate you.

Thanks as well to Ashley Pappas, Tamara Harris, and Xenia Wolkoff of Industry Wizards, for their creative suggestions and their diligence and hard work in creating the community of professionals represented in *Thin Air*.

Thanks, of course, to Sara Maurno for many months of seemingly limitless patience; and Sofia Bella Sirico, whose inspiration spurred her daddy onward.

And finally, humble thanks to those who shall remain nameless: representatives of major companies who were willing to share their wisdom but unable to disclose their identities for competitive reasons. "There're just a couple of us in this field," as one interviewee put it, "so we don't give anything away. But this is too important to business to say nothing." Indeed it is.

# Introduction—
# The Confusion of the Tongues

Remember the biblical story of the "Tower of Babel?"

The short of it was that the Almighty, annoyed by the presumptuous Babylonians, gave each of them a different language. Thus, two friends (perhaps fishermen who owned a boat together) might have arrived to watch the tower being built, but found themselves suddenly unable to understand each other. Perhaps one spoke Portuguese, and the other, Mandarin Chinese. We can picture our two friends, unable to understand each other, arguing in vain over who would get the boat. But rather than learn each other's tongue for their mutual benefit, they scattered.

The Lean and Wireless camps are something like those fishermen. They have essentially the same mission: to enable productive and profitable enterprises. But they do not speak one another's tongue. A Six Sigma Black Belt we interviewed for *Thin Air* asked, "What in hell is 'weefee'?" He was referring to Wi-Fi, which is an established and rapidly evolving standard for enterprise telephony and networking (and which enables students on college campuses and patrons at Starbucks to connect to the Internet at any time). An equally successful VP of business development from a Wireless technology provider asked in a note, "What is a 'can ban'?" He was referring to *kanban*, the Lean inventory tool, and he also described Lean as "that fad from the '80s" and wondered if anyone still bothered with it.

Only the rare few speak both tongues. The bilingual include enterprises such as Boeing, the Mercy Medical catheter laboratory in Des Moines, Iowa, and Gulf States Toyota, an independent distributorship out of Houston, Texas. In fact, the project manager for a Wireless implementation at Mercy Medical is a Six Sigma Black Belt. On the other side of the fence, a select few of the Wireless technology providers, from companies such as Motorola, Ekahau, Zebra, Omnitrol Networks, and Rush Tracking Systems, are led by men and women who know the language of continuous improvement very well. Motorola, for example, acquired Symbol Technologies, which won the Shingo Prize for Excellence in Manufacturing in 2003.

As a whole, though, the Wireless camp has never heard of kanban, *kaizen* (which engages those who use a process to improve and standardize it), or *poka-yoke* (the practice of mistake-proofing a process). It's not that the Wireless camp disapproves, but approaches those challenges differently. For example, although they don't call it *poka-yoke*, Alameda County in California uses radio-frequency identification (RFID) technology to mistake-proof the collection of polling equipment (and, subsequently, votes).

The Lean camp tends to hold fast to its commitment to "rules, not tools"—that is to say, process improvements over technology—in continuous improvement. In practice, Lean practitioners are not averse to tools; electronic kanban or eKanban is emerging as a norm. But these experts as a whole do not partner with technology providers, nor do they involve themselves in technology implementations. In essence, they help to create a system and culture of continuous improvement, then move on to the next engagement. But of course, the need for improvement never ceases; it just moves around the enterprise.

The list of Wireless technology providers is growing, and includes equipment providers such as Motorola, system integrators such as Rush Tracking Systems, and enterprise companies such as SAP. Lean consultants have partners of their own, but the two camps (Wireless and Lean consultants) have no partnerships between them. Despite the fact that both camps are accomplished in enabling enterprises to improve operations and profit, neither particularly understands nor trusts the other's school of thought.

This is a shame, because the two camps have the same objectives (to remove waste, to standardize processes, to reduce costs, etc.) and each has proven methodologies to achieve those objectives. But neither approach is enough by itself: Ford, for example, is an absolute champion of automated process efficiency (using RFID in particular), and Toyota pioneered continuous improvement, but neither weathered the 2009 recession particularly well. Companies that went gung-ho for Six Sigma found themselves hard pressed to make the extensive measurements required by Six Sigma, and Six Sigma supposedly stifled innovation at 3M. Without the automated measurement that Wireless technologies provide, Six Sigma was simply impractical. The combined value proposition of Lean (or continuous improvement) and Wireless is greater than each by itself.

In this book, we propose some practices and paradigms that create that single, stronger value proposition, marrying these two methodologies. This calls for a common language.

Throughout this book, we use the capitalized term "Wireless" to describe a school of numerous and disparate technologies and the discipline surrounding those technologies.

Wireless encompasses two broad classes of technology. The first is *Wireless Communication* (smartphones, cellular phones, PDAs, and wireless computers). The second is *Tactical Wireless* (radio frequency identification [RFID], real-time location systems [RTLS], global positioning systems [GPS], remote sensors, etc.).

We also touch on "Airsourcing," the Wireless cousin of outsourcing. Outsourcing does not eliminate work, rather, it commissions it from elsewhere, from someone. Similarly, Wireless does not eliminate work, it assigns it to some*thing.* Outsourcing is macro, referring to a product design, or its manufacture, or the like. Airsourcing is micro and task oriented; it performs tasks, such as conveying tire pressure from the tire to an onboard computer, or generating an advanced shipping notice for a pallet of goods that was just loaded onto a truck. These useful tasks are not eliminated, but the speed at which they are performed is increased; and in many cases, the labor required to perform the task is eliminated or decreased. Wireless technology automation replaces or reduces the labor.

We use the term "Lean" broadly, to refer to systems and tools (such as 5S,* kanban, just-in-time [JIT], and kaizen) that as the Lean Learning Center describes, "[give] people at all levels of the organization the skills and a shared way of thinking to systematically drive out waste through designing and improving work of activities, connections, and flows,"[1] which is exactly the mission of Wireless, as well.

## THE LEAN WIRELESS ROI CALCULATOR

You are eager to understand just how Wireless can return money to your organization (hence, your purchase of *Thin Air*). With that purchase, you are entitled to use the Lean Wireless ROI Calculator, a one-of-a-kind resource for companies considering wireless technology. (Think of

---

* Sift, Sweep, Sort, Sanitize, and Sustain, a methodology that organizes what we need to complete a process and eliminates what we do not need, thus allowing us to identify problems quickly.

it as a sort of Kiplinger TaxCut® of Wireless; it asks you for some figures, then performs complex calculations.) Block off perhaps a half hour of your time, visit www.leanwireless.com, and see what Wireless can do for you.

## "E" BEFORE "I"

The "e" in "e-mail" and "eKanban" stands for "electronic."
The "i" in "iPod" and "iPhone" (both Apple products) stands for "Internet."

The Lean camp is perfectly familiar with the term eKanban. But, as Wireless continues to connect over the Internet, intranets, and virtual private networks, and as businesses use the Internet to collaborate remotely, perhaps iKanban is a more proper term. Thus, we refer in *Thin Air* to iKaizen, iPoka-yoke, and so on. (If Apple has trademarked the letter "i," we are unaware of it.)

## DEMOCRATIZATION

We refer often to democratization in *Thin Air*. Lean was born in manufacturing, and manufacturers typically look to other manufacturers for best practices. Not so in Wireless. By democratization, we mean that the best practices in Wireless are found:

- In all industries
- In nonindustry (chiefly, consumer uses)
- In all employee levels

There are plenty of instances of manufacturers using Wireless to improve processes, and there are also plenty of success stories from the government, service, retail, and education sectors. The short list of nonmanufacturer enterprises we cover in later chapters includes

- The U.S. Air Force
- Mercy Medical Center's catheterization laboratory
- The Thames Valley Police (United Kingdom) and the Cape Breton Regional Police Force (Canada)
- The Florida Court System
- Alameda County (California) Registrar of Voters
- Numerous libraries and university systems

Moreover, we show how the "consumer effect," the demand for always-on and ubiquitous connectivity, is shaping the Wireless enterprise. For example, consumers demand the ability to download the latest hilarious video from YouTube onto their smartphones; their demand has driven the evolution of high-quality high-bit-rate streaming media. Industry can use it, too, and industry does. Finally, consumers with their plug-and-play sensibilities enter the workforce and demand to know why they can get a wireless connection at a coffee shop, but not in a meeting room. Or why every device that they plug into their computers (be it a mouse, headset, gaming platform) plugs into a USB port, whereas Wireless sensors don't. Good questions.

And so advances in Wireless are not confined to a given industry, and not driven solely by management; in essence, we all spur Wireless onward.

Conversely, Lean methodologies, after nearly two decades, have barely stretched themselves beyond manufacturing. You hear about the Lean factory, but not the Lean university. You hear about the Lean supply chain, but it's still largely an ideal. Occasionally, you'll hear of Lean government (the Environmental Protection Agency and the Department of Defense [DoD] in particular have embraced Lean practices), but government as a whole has never heard of Lean.

And manufacturers themselves embrace Wireless to improve their processes. Industry consultants ARC Advisory Group's 2008 market study "Wireless Technology in Process Manufacturing Worldwide Outlook" forecast a compound annual growth rate of 30 percent among process manufacturers; manufacturing, ARC concluded, is rapidly adopting the Wireless technologies that have been developed for the IT, telecom, consumer, or military markets.[2] Wireless technology providers have broad portfolios, reaching into verticals (such as government, education, and retail) where continuous improvement experts have yet to go. But as we explore in *Thin*

*Air,* those experts in Lean and continuous improvement have much to contribute to Wireless success.

## TECHNOLOGY AND THE CLASH OF CULTURES

A recurring theme throughout *Thin Air* is the clash that occurs when one group is asked to change its thinking (be it the Lean camp, Wireless camp, management, young workers, etc.). The Lean camp is fond of saying that "rules, not tools" foster continuous improvement. We undertook *Thin Air* to propose a new paradigm, "rules *before* tools," and to foster partnership between the Lean and Wireless camps. We believe that a new subset of Lean, *Lean Wireless,* is absolutely necessary, a focused practice that combines the strengths of Lean and Wireless, and mitigates the weaknesses. Wireless will advance, but must not advance without regard for profit, added value, or security; here is where the technology requires Lean expertise. We hope that each camp will forgive the sometimes basic information peppered throughout these pages, but some of the concepts and terminology will be new to either camp.

We also undertook *Thin Air* to examine the sometimes troublesome but meteoric rise of Wireless and Web 2.0 technology, and to suggest methods that the Lean and Wireless camps can use to smooth the transition.

A long-held tenet of anthropology is this: no culture ever met a more technologically advanced culture to its own benefit; which is, pretty much, what is happening in business, as Baby Boomers and middle-aged workers meet the Millennials, workers aged 27 and under. The Millennials, sometimes called "digital natives," grew up online, adopting whatever latest technology or social application existed. Now, they're bringing their skills and sensibilities into the workplace.

Usually, the less advanced of two cultures holds itself in high esteem for its ingenuity, fighting prowess, healing powers, what have you. Once it discovers that greater technology exists, it loses confidence—in its leaders, religious figures, military, and art—and the culture, in essence, dies. This supposedly contributed to the demise of the Neanderthal, who never conceived of the throwing spear until they met the Cro-Magnon, and of the Maya, who until they met the Spanish, had not conceived of gunpowder and metallurgy. The Maya ceased to worship Kukulcan because the European god seemed more generous and powerful.

Consumers are no different. CDs quickly put vinyl records to death, and although television did not kill radio, it quickly dominated it. Under no circumstance will consumers accept lesser technology and capabilities in their workplace than they enjoy outside of it. Certainly a company can take the stance that "the employee doesn't get a vote," but such a company is simply noncompetitive, even quaint, in hiring talent. It is foolhardy to resist Wireless and Web 2.0. The digital natives have seen it, and their culture is inextricably transformed.

To understand just how much Wireless affects every working American now—and increases connectivity and productivity—observe the day-in-the-life of an American worker:

- For better or for worse, he gets his news not from a newspaper, not even the radio any longer, but the Internet.
- He receives a traffic alert wirelessly on his car's dashboard, and is able to circumvent a traffic jam.
- He sails through a toll plaza without stopping, using his E-ZPass.
- He fills up his gas tank without having to pull out his wallet or spend time giving cash or a credit card to a cashier. It is all done via a contactless payment system, on a key fob or on his dashboard.
- He takes a call on a cell phone from an underling who is ill, and phones a foreman to reallocate workers before the shift begins.
- At work, he takes his laptop computer into a meeting, and connects to the network and then to a presentation on a server over a wireless local area network, or WLAN.
- He connects to a remote colleague using a voice-over-wireless local area network, or VoWLAN; they see one another with a video-over-Internet application like Skype.

The Wireless society is already here, or at least, is growing deep roots.

But once again, it is vital that Wireless enables and improves business, which it does, when used well. Wireless must not be allowed to distract employees from their core missions, expose enterprises to new security risks, or create new forms of waste, which it also does.

If you are a Wireless technology provider (or user), then *Thin Air* will help you to understand the impact of Wireless on your organization, and help you to keep it focused on creating value. If you are a Lean or continuous improvement practitioner or consultant, *Thin Air* will help you recognize how the technology fulfills the Lean mission, and how you can plug it in to your strategies.

## ENDNOTES

1. Flinchbaugh, Jamie. 2009 *Beyond Lean: Building Sustainable Business and People Success through New Ways of Thinking.* Novi, MI. ©Lean Learning Center.
2. http://38.119.46.100/fieldserviceusaeast/whitepaper.pdf

# 1

## *Lean Wireless Is Already Here*

If you're far along enough in your career to be responsible for networks, infrastructure, or operations, you're probably not wearing a fine jersey muscle T-shirt or a pair of killer pink leggings, and you may have never set foot in an American Apparel retail store. But you should do exactly that. American Apparel is both a Lean operation and a Wireless one, and it traces a 14 percent jump in sales at seven stores directly to using RFID Next. To any competitor that targets young people (such as Abercrombie & Fitch or Old Navy), an American Apparel store seems downright sparse in inventory (see Figure 1.1). But it is not sparse at all; rather, it is precise in its inventory. There is exactly one of each item in a given style, size, and color on the rack at any one time.

In its Columbia University location in New York City, a young woman selected a pair of buttercup-yellow leggings (in an improbably small size) and made her way to the register. After she'd paid, but before she had left the store, another young woman had replaced the leggings on the rack with an identical pair, after receiving a signal on a PC that those exact leggings needed replenishment.

That precision goes beyond American Apparel's brick-and-mortar stores to its Web page; there, each section (Men, Women, Kids, Babies) lists the top ten sellers: "This week's most popular products based on what's been getting your attention and flying off our shelves. These styles are the most looked-at and most purchased products from the last seven days."[1] Check back weekly, American Apparel advises, as it updates its sales statistics every Monday on "what's hot."

This is a remarkable example of Lean Wireless: the store uses radio-frequency identification (RFID) to maintain a precise inventory and satisfy its patrons, and Web 2.0 to connect its customers to its production lines in

**FIGURE 1.1**
Room to breathe, and no stock-outs. RFID at American Apparel stores frees workers from manual inventories, and ensures that customers can buy just what they want. (Courtesy Motorola, Inc.)

pull-from-demand relationships and to gather rich informatics that it uses in sales promotions.

Lean Wireless is proven at American Apparel, and anything that's proven in fashion retail is well-proven indeed. Success and profit in clothing retail depend upon putting the right item in the right style and size in the customers' hands without fail, or they'll go elsewhere.[2]

One of the biggest obstacles to customer satisfaction is the customers themselves, who move stock from its location, leave it in the dressing room, and, on occasion, steal it. Theft and false stock-outs are not merely the "costs of doing business" for retailers. The Harvard Business School estimates that eight percent of all retail items are out of stock at any one time, which costs the top 100 retailers about $69 billion annually.[3] A similar survey in 2006 by the University of Florida discovered that retailers lost more than $41.6 billion (1.6 percent of overall sales) to theft. Between all those factors, the Harvard study found up to 65 percent inaccuracy in inventory counting.[4]

American Apparel faced the same challenges. At any one time, says American Apparel's Research and Development Strategist, Zander Livingston, "We had 10 or 20 percent of inventory lost throughout the stockrooms. My goal was to locate these items and take them to the sales floor. We solved that with RFID."[5]

The process begins at the manufacturing plant in Los Angeles. Here, employees affix Avery Dennison AD-222 RFID tags to each clothing item's price tag. The RFID-enabled stores use four RFID stations fitted with Motorola XR-Series RFID readers:

1. A *receiving station*, at which employees use RFID to enter incoming shipments.

2. A *fill station* in the stock room, which informs employees onscreen what items/colors/sizes have sold and must be replenished.
3. A *validation point*, between the stock room and sales floor. Here, the employee waves the items past a stationary reader, which ensures that the employee is replenishing the correct items.
4. A *point-of-sale station* at the cash register. This immediately alerts employees at the fill station described in Step 2.

Replenishment is ongoing and ceaseless. As a young woman said in an Avery Dennison promotional video, "If something gets sold and is no longer on the floor, it's usually no longer than 30 seconds to a minute before it's back on the floor."[6]

Still, American Apparel double-checks itself by taking a storewide inventory with a handheld Motorola MC9090-G RFID reader. This inventory, according to Livingston, takes one employee two hours. Without RFID, the same task would take five employees five hours.

American Apparel piloted RFID at its Columbia University store in New York City. The system was largely plug-and-play, and Livingston estimates that it took him and an assistant an hour and a half to implement RFID at that store, and at each of six other New York City locations.

The Lean benefits to this approach were nothing short of remarkable. At seven RFID-enabled stores, American Apparel reduced labor by an average of 60 hours per week in each store (and 100 hours in one of the stores). At the same time, inventory accuracy increased to an average of higher than 99 percent. And, says Livingston, "We've seen a 15 percent sales lift across our seven stores, compared to the rest of the fleet. And you see results like that across the board. Walmart tagged one item, an air freshener, and saw a 25 percent sales lift." The day we spoke, Livingston had just received approval to implement RFID at 19 more New York City locations. Something the 19-store roll-out will incorporate is loss prevention, using RFID at the exits to alert staff of theft, which will further increase the return on investment.

A final benefit to American Apparel's Lean Wireless approach is that it empowers employees. American Apparel has a comparatively young staff working in its stores, between 18 and 22 years of age, and Livingston believes that eliminating the routine tasks of counting clothes and conducting before- and after-hours inventory is a differentiator in attracting top talent. "It lets you spend more time with customers," said one young woman, and a young man remarked that he would "never go back to the old system. Everyone should have this."[7] As retail jobs go, young people find American Apparel jobs to involve far less of the usual inventory drudgery.

In short, RFID enables American Apparel to create rules for its stores, and to enforce those rules with Wireless tools.

## RULES BEFORE TOOLS

As we described in the introduction, Lean theorists favor business rules over technology tools, improving a faulty process over simply implementing some technology and expecting it to solve the fault.

As a whole, the technology camp agrees. SAP CEO Claus Heinrich wrote in his 2005 book, *RFID and Beyond,* that "Implementing just IT systems without their supporting business processes is a waste of money."[8]

In his book, Heinrich cites research conducted by SAP and PRTM Management Consultants of 60 companies and 75 supply chains. As a whole, companies with mature business processes but outdated IT had lower inventory-carrying costs and higher profit than companies with best-in-class IT configurations (and presumably, bad business processes). Technology by itself did not fix what was broken, but seemingly helped to make more mistakes, and to make them faster.

Wireless tools such as RFID are just that: tools. The same goes for the Lean tools of kanban and kaizen. But Lean consultants generally steer clear of making technology recommendations for achieving continuous improvement. As Jamie Flinchbaugh, co-founder of the Lean Learning Center, explains this dichotomy, "I don't think the Wireless integrators and Lean consultants have the same goals. Lean is selling an *approach* to working on problems and improvement. Wireless integrators are selling a specific set of solutions for a specific set of problems." Lean consultants accept that technology makes some improvements possible, but as a whole, do not believe that a lack of technology constrains improvement.

Still, the Lean and Wireless camps have more in common than they recognize. They use different terms—Lean refers to the *gemba*, where value is created, whereas Wireless refers to "the edge," where work is performed—but these are just semantics. Both camps concern themselves with improving actual processes. But American Apparel did not conduct a kaizen to combat theft; it installed RFID and solved an inventory challenge. Interestingly, the PRTM study to which SAP's Heinrich referred revealed that companies with mature business processes have 28 percent lower inventory levels than those who rely upon technology alone, a remarkable Lean achievement. Lower inventory represents lower costs of storage and fewer losses due to obsolescence, and an accurate inventory means that there's no need for "buffer stock" to make up for lost or misplaced inventory. Couple mature processes with Lean Wireless technology, as American Apparel has, and a company has its best chance of approaching 100 percent inventory accuracy.

It is a mistake to avoid technology tools; trying to be a competitive retailer without RFID inventory or security tags or contactless payment is like trying to chop wood with an ax. The guy with a chainsaw will chop far more wood in a day, and he'll become the go-to guy for cordwood.

## Complex Business Environments Require Technology for Visibility

In *RFID and Beyond*, Heinrich introduced the concept of "real world awareness," which goes beyond real-time data or real-time reporting. The primary principle underlying real world awareness is the need to act on facts, rather than upon assumptions or historic data. As Heinrich describes it, "Automated systems can sense the exact state of the real world and respond appropriately."[9]

This is akin to the Lean principle of reacting to an actual order, and the constraints-based principle of managing an existing process bottleneck to increase overall throughput (and then looking for the new bottleneck).

In a complex environment, real-time response and managing constraints are impossible without technology. Heinrich detailed the example of an airplane, in which sensors track real-time parameters such as speed, altitude, fuel levels, and the like. Over time, aircraft that once relied upon a pilot's instinct (such as the WWI Sopwith Camel or Fokker Triplane) evolved through complex instrumentation and improved processes to require a co-pilot, navigator, radio officer, and flight engineer; then, through automation, the navigator, radio officer, and flight engineer were eliminated, and pilots and co-pilots managed by exception (when some parameter was out of the ordinary) and insight—the Lean way.[10]

The customer is the passenger, who sees none of these complex mechanics first hand. The passenger sees the results in safety, affordability, and "on-time delivery," and is presumably satisfied.

Technology expands the reach of Lean, believes Kevin Prouty, who was the senior director of manufacturing solutions at Motorola. Prouty, now with Infor Global Solutions, is a luminary in both Lean and Wireless theory, having been with Symbol Technologies when it was acquired by Motorola for Symbol's manufacturing RFID solutions. Symbol brought to Motorola its strong Lean sensibilities, having won the prestigious Shingo Prize for Excellence in Manufacturing in 2003. Symbol's then-president, Dr. Satya Sharma, had previously led Lucent to become the first company to win Japan's Deming Prize, and was a firm believer in Lean. Under Sharma, Symbol implemented Lean practices such as poka-yoke companywide in its manufacture of Wireless technology.

"I always believe that you Lean your process," says Prouty, "and then start filling in the gaps in technology. The reason I say that is, poka-yokes and kanban can only go so far manually. Kanban is a great process by itself, but a local process. If you try to extend kanban beyond a card system, you need some technology to transmit beyond the local area," for automatic alerts to material handlers, or to involve suppliers. This is why eKanban is the first well-accepted Lean Wireless practice. Most large automotive companies, for example, use some form of eKanban.

## Lean Leaders Embrace Wireless

Lean is based largely on the Toyota Production System (TPS), which Toyota used to produce the lower-cost, higher-quality automobiles that took U.S. automakers by surprise beginning in the 1970s. It is because of Lean's roots in TPS that Lean uses such Japanese words as *kanban*, *kaizen*, *gemba*, and *poka-yoke*.

If Toyota is responsible for Lean, then, does Toyota believe in rules, not tools? Hardly: Toyota makes good use of tools but does so in a Lean fashion. The Toyota plant in Kentucky is one of the largest auto plants in the world, comprised of more than 7.5 million square feet, and containing more than 1,200 robots. Management uses the traditional highly visible Andon boards (common among Lean manufacturers) which display key performance indicators (KPIs) in central locations, but also electronic Andon boards that it pushes out to employees through the local area network (LAN) and wirelessly to their BlackBerry™ devices. This was the only way to make Andon boards visible to every employee working within those 7.5 million square feet.[11] For its implementation in maintenance, the Kentucky plant involved six key maintenance personnel in creating screens and processes for mobile devices by which they could monitor critical equipment (a utility we call iGemba) in real time. Toyota estimates that this utility alone saves the plant $1 million per year, through more immediate response to equipment or other malfunctions.[12]

Toyota's Taiichi Ohno is credited with creating the TPS, which he detailed in his book *Toyota Production System, Beyond Large-Scale Production*. But as industry sage Ralph Rio of ARC Advisory Group put it, Ohno released the book in 1978, in the days of batch processing, voluminous print reports, and even punch cards. "Ohno-san was right when he advised companies to avoid technology," said Rio, but, "It is an understatement to say that a lot has changed in the thirty years since 1978."[13]

Entities such as Walmart, Boeing, and the U.S. Department of Defense (DoD) are deeply invested in both Lean and Wireless practices. Walmart may not call it kanban, but its requirement that its suppliers place RFID tags on inbound shipments is aimed at maintaining inventory, enabling pull-from-demand, eliminating stock-outs, and reducing labor: all Lean ideals.

The A3 form (so named for a common paper size used in Japan) is another Lean tool, one that employees use to make a suggestion for improvement. Boeing does not call it A3, but has introduced a lightning-fast A3 practice at its Seattle production plant, where the 787 Dreamliner is manufactured. Production workers armed with BlackBerry devices are able to request and receive permission for engineering changes on the spot. (This is an enormous leap in efficiency: in the 1980s, co-author Dann Maurno worked for a defense contractor, where a simple engineering change such as repositioning a graphic on a missile chassis could take weeks to enact.)

Wireless technology is not limited to production at Boeing, but is also embedded in the design of the 787 Dreamliner itself. At any one time, the Dreamliner in operation has 48,000 data points encompassing structural health and aircraft prognostics, among other indicators.

The Department of Defense sees Lean and Wireless as interrelated. The United States Air Force's eLog21 is its supply-chain transformation aimed at delivering warfighters to engagements rapidly and effectively.

The USAF looked to the private sector for best practices in the supply chain to deliver what it called "overwhelming superiority at an affordable price." Among other aggressive measures, it "Leaned" its IT backbone, halting upgrades and development on 400 of 1,100 legacy applications, which freed $347 million to upgrade mission-critical applications; it created a Lean supplier base, including a single systems integrator; and it made ample use of RFID as a Lean logistics tool. Since 2003, the DoD has required its suppliers to use RFID on packaged products, and it uses portable RFID on the ground in Iraq and Afghanistan to process logistics and movements.

Finally, enterprise software giant SAP AG achieved the Lean ideal of flexible configurations at its Walldorf, Germany headquarters. There, the company uses more than 2,000 wireless light and blind controllers manufactured by EnOcean, a Siemens AG spinoff. In addition, because the SAP building has no fixed inner walls and requires flexible room structures, the company can change those structures without building walls or undertaking the costly rewiring and rerouting of cables.

## To Ignore Wireless Tools Is to Forgo Progress

As we described earlier, Wireless does not contend with or supplant Lean; it extends its reach in a huge enterprise such as Toyota, achieves Lean results in companies such as American Apparel, and enables the sustainable improvement to which Lean aspires. Thus, for companies looking to drive down their costs of doing business, we propose a new paradigm of "rules before tools = results."

"To ignore the available tools is simply to limit progress," said Ravi Pappu, co-founder of Cambridge, Massachusetts-based ThingMagic. "I take issue with slogans like 'rules, not tools' and 'drill baby, drill,' because at best they rule out interesting possibilities." Pappu described the birth of the Internet, which was conceived by the U.S. Defense Advanced Research Projects Agency (DARPA) to trade information among a few key researchers. "If you told them 'you need the interoffice communication envelope instead of the Internet because only four of you want to send data back and forth,' then we'd still be filling out little boxes on envelopes. And everyone will agree it is pretty convenient to buy a book on Amazon when you need it. There is a time for sloganeering, and innovation is not that time."

## Wireless in Lean and Six Sigma

The missions of Lean and Six Sigma have always been missions of continuous improvement and process improvement. Wireless may not have started out as a process improvement tool, but that is the direction it is taking now and it should be part of the Lean and Six Sigma "toolkit."

"When I first began Omnitrol, a 90 percent driver for the auto ID [automatic identification] market was tracking goods," said CEO Raj Saksena of Omnitrol Networks. "What I've seen in the course of time is that suddenly, *improving internal processes and operations* has become the driving factor for interest in RFID technology." Tracking goods is still on the table, of course, but Saksena observes a real interest in using RFID for Lean objectives, including:

- Gathering production informatics
- Tracking work-in-process and employee productivity
- Identifying bottlenecks or proactively staving off bottlenecks

"That is how the market perceives the technology now," said Saksena.

Rio of ARC Advisory Group described five ways in which Wireless Communication is consistent with Lean and Six Sigma methodologies (see Table 1.1).[14]

Wireless:

- Connects people to the process so that "go and see" (called *Genchi Gembutsu*) includes mobile employees*
- Enables fact-based decision making and real-time decision support
- Includes the entire team, both stationary and mobile workers, in making decisions by consensus
- Gives everyone access to the same data or "one version of the truth"
- Provides "visual controls" (mobile Andon and LCD displays) through mobile devices, particularly to support engineers and management

Both Lean and Wireless are largely democratic. They are driven and perfected by the crowd. "Everybody in Lean is responsible for continuous improvement," said Rio. "It isn't just some chieftains who are anointed

---

* Genchi Gembutsu is somewhat a catch-all term, but it involves sending continuous improvement teams to observe and troubleshoot a process firsthand.

**TABLE 1.1**

Comparing Lean Manufacturing versus Six Sigma and the Wireless Benefits to Each

| | Lean Manufacturing | Six Sigma |
|---|---|---|
| Purpose | Reduce Waste | Reduce Variation |
| Focus | Value Stream | Each Operation |
| Methodology | ID Customer Value | Define |
| | Value Stream Map | Measure |
| | Flow | Analyze |
| | Pull | Improve |
| | Perfection | Control |
| Penetration in 2007 | 69% | 52% |
| Wireless Benefits | Real-time information from ERP, MES, Quality, Maintenance, Automation, and other systems for fact-based decision making | |
| | Real-time visibility to know when to "Go and See." Review non-value-add steps in value streams for new opportunities to remove waste. Improve mobile employees' involvement particularly for standardized work and sharing best practices. Without hardwired network connections, improved flexibility for kaizen teams to move equipment for flow improvement. | Improve capability to collect data for measure phase in the DMAIC methodology. With more flexibility to locate sensors for process monitoring, improve the control phase. Improve process control and reliability with the capability to eliminate wiring and put sensors on mobile parts of equipment. |

*Source:* ARC Advisory Group.

to be the continuous improvement people." Similarly, mobile computing (using wireless notebook computers, tablet computers, and application-rich PDAs) is democratic, allowing participation by everyone within an organization, with the same tools.

Wireless enables Six Sigma's DMAIC process of define/measure/analyze/improve/control. Specifically, DMAIC means:

- *Define* the current process, and goals of the improvement.
- *Measure* the current process, objectively.

- *Analyze* the data, looking for cause-and-effect relationships.
- *Improve* the process based on the measurement and analysis.
- *Control* the process, such that it is consistent and without defects. Control typically involves control mechanisms and continuous monitoring.

For those unfamiliar with Six Sigma, the sigma level refers to percentage of defects versus percentage of yield. A sigma level of 1 (One Sigma) produces 69 percent defects, thus a 31 percent yield. Pretty bad, no matter what the process. A sigma level of 2 represents 31 percent defects, 69 percent yield; and so on to Six Sigma, of .00034 percent defects, 99.99966 percent yield. Sigma levels vary in criticality, depending upon the process. If the process is manufacturing a dress-up doll for kids, then perhaps Four Sigma (with percent defects of .62 percent) is good enough. In that case, improving the process can cost more than the defects. But on the other hand, if the process is a safe flight from New York to Los Angeles, then Four Sigma would mean that one out of 100 flights would fail to reach Los Angeles, which is of course unacceptable.

Wireless enables several of the steps of DMAIC. First, Wireless sensors and monitors can measure the current process. Automated measurement is ideal in that it does not disrupt the existing process, does not involve labor, and can make thousands of automated measurements without error. Manual data collection typically has an an error rate of 2.9 Sigma (or percent yields just under 93 percent). As Rio describes the need, "Using manual data collection in the assessment of a Three Sigma or higher process *produces more errors than the process itself.*"* That is, unless defects exceed seven percent of a process, manual measurement may not enable any improvement. Practically speaking, DMAIC requires automated data collection to achieve Three Sigma or higher. Wireless sensors may also be used on remote or moving equipment, where it is impractical for humans to take measurements.

If there is any doubt that Lean and Wireless missions are largely the same, take a look at Figure 1.2. This methodology is used by Rush Tracking Systems in designing systems for numerous clients, including the U.S. Department of Energy and American Dairy Brands. Although the methodology appears complex, it reflects the company's (and CEO Toby Rush's) mastery of Lean methodologies. It begins with the enterprisewide value

---

* Italics ours.

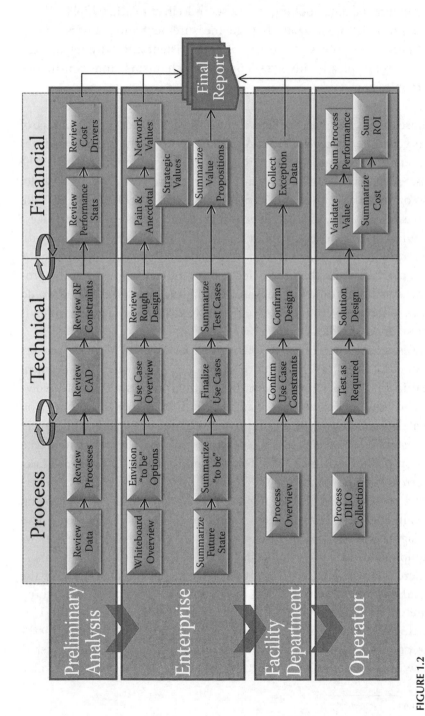

**FIGURE 1.2**

Enterprise RFID assessment methodology. (Courtesy Rush Tracking Systems)

streams, then moves downward to the department level, and then seeks intelligence from the edge, at the worker level. Observe the number of steps that hint at Lean and continuous improvement methodologies·

- Envision "to be" options
- Process overview
- Review RF constraints
- Review performance statistics

It is no accident that Wireless, which was once an information technology or IT priority, increasingly rests with Operations and as the purview of the Chief Operating Officer, which is also where Lean and Six Sigma priorities reside.

## THE MARRIAGE OF LEAN AND SIX SIGMA

Continuous improvement, as a discipline, has continuously evolved. Lean of course focuses upon eliminating waste, using specific methodologies to help identify where the waste is. Six Sigma aims to eliminate variation in processes or products. Here's a simple example of Six Sigma for our Wireless readers. If you are cutting some kind of metal blank to a length of one inch, and specifications allow a tolerance of ±30/1000ths of an inch, you aim for a tolerance of ±5/1000ths. By doing so, you are less likely to produce something outside of the looser spec, and will produce to the Six Sigma level of 3.4 defects/million. By "shooting for the stars," you reach the moon.

Although Lean and Six Sigma are both methods of continuous improvement, Lean appears to have succeeded where Six Sigma has failed. If it was not dealt a death blow in 2004, it received a resounding slap from Jay Desai, a corporate strategist for General Electric. Prior to 2004, GE had famously promoted Six Sigma for nearly two decades, but Desai said in a Reuter's news report that Six Sigma's narrow focus on discrete problems was slow, and likely constrained innovation. ("If Lucent applies Six Sigma," Desai speculated, "they die."[15] Lucent, at the time, was attempting to recover from $30 billion in losses.)

Our view is that Six Sigma has not failed, but has moved under the Lean umbrella. The two have been morphing for years, and "Lean Six Sigma" is now a fairly common term. Never mind that the methods are different: users perceive them as the same and use them together. Users will always defy design and do with a tool as they see fit, a theme we return to several times in *Thin Air*.

This marriage was a natural and welcome progression. Lean is an organizational discipline, whereas Six Sigma is a point solution (solving localized problems), and they complement each other marvelously. Six Sigma brings to the marriage a healthy respect for tools, analytics, metrics, and efficiencies; Lean brings its strong focus upon the value stream. Six Sigma favors an elite few who are solving a given challenge, whereas Lean is democratic, involving everyone in an organization. As we show throughout *Thin Air*, democracy is powerful, and has driven the meteoric advance of Wireless technology (for both good and ill).

## LEAN'S NEW NEIGHBOR—WIRELESS

If we think of Lean and Six Sigma as the "Continuous Improvement Family," they have a new neighbor, Wireless. Wireless is somewhere between a discipline like Lean, and a point solution like Six Sigma. To extend the family metaphor, think of the Wireless family as consisting of Tactical Wireless and Wireless Communication. Tactical Wireless is largely invisible to business and consumers, and so is less well known. It includes such technologies as radio frequency identification (RFID), real-time location systems (RTLS), wireless sensor networks (WSNs), and global positioning systems (GPS), among other technologies. Wireless Communication, on the other hand, includes such recognizable form factors as Wi-Fi-enabled laptops, PDAs, smartphones, and cellular phones.

The Wireless family also includes a gifted but sometimes bratty teenager named Web 2.0. That name incorporates some seemingly trivial applications such as social networks (Facebook, MySpace, YouTube) and online gaming. It also includes such business-oriented applications as 24-hour Internet connectivity, rich Web interfaces (including streaming video and Web conferencing), and e-mail marketing. Web 2.0 works hard, plays hard, and frequently plays at work.

If we think of them as a family, Tactical Wireless and Wireless Communication are becoming inseparable, and they always seem to bring along the kid; for example, the wireless sensor on a machine (a Tactical

Wireless application) can send a signal to a machine operator on a PDA (a Mobile Communication device). Or, the RFID-enabled smart shelf (Tactical Wireless) at a hospital might signal a supplier for replenishment, through a virtual private network (Web 2.0).

## Wireless on the Value Stream

Lean experts will find that Wireless providers understand value streams surprisingly well; they simply do not call them "value streams." Case in point? Omnitrol Networks, which provides a Smart Infrastructure Service Emulator, or SE (see Figure 1.3). This creates a virtual Wireless environment to test real shop-floor devices and business services, and to perfect their use. The emulation is, for all practical purposes, a value stream diagram.

Using emulators and virtual applications like them, Wireless providers can achieve what Lean experts aspire to achieve, including:

- Reducing risks and costs, in technology deployments
- Testing theoretical load levels (to identify bottlenecks)
- Sharing success enterprisewide, with collaborative Web-based development and test environments
- Creating feedback loops, through record and playback functions for systematic testing
- Troubleshooting processes and workflows, in both simulated and real-life configurations
- Labor savings, through remote testing and monitoring via a Web-based interface

Emulators and simulators can save enormous amounts of time and money; they allow companies to simulate how a new device or set of devices will interact and perform before deciding to purchase that equipment, and without interrupting production. Emulators do not create a direct value to customers, but the customer sees the benefits in the delivery and cost of goods sold.

Thus, Lean practitioners will find that Wireless tools such as these offer a Lean value proposition. Increasingly, when enterprises seek process improvements, Lean consultants and integrators will find themselves in direct competition with Wireless technology providers, or better still, working alongside them.

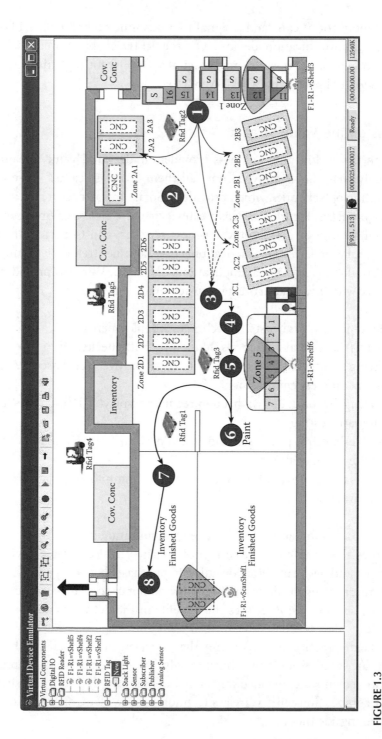

**FIGURE 1.3**

A wireless service emulator. It creates what is, in essence, a value-stream diagram for Wireless nodes. (Courtesy Omnitrol Networks, Inc.)

## THE WIRELESS TSUNAMI

With all of the drivers behind the Wireless momentum from consumers, government, industry, and retail, the momentum appears unstoppable. Set aside the PDAs and wireless devices for a moment, and consider the most utilitarian Wireless application of all: RFID. The worldwide RFID market exceeded $5.3 billion in 2008, according to industry think tank ABI Research, with short-term growth driven by such high-growth, high-volume applications as supply chain management, ID documents, ticketing, and contactless payment.[16]

In 2008 alone:

- Retailers American Apparel, New Balance, and Jones Apparel/Nine West all moved to item-level RFID tagging (long considered cost prohibitive) to ensure product availability for improved customer service.
- The state of Hawaii used RFID to guarantee the quality and safe consumption of agricultural products.
- Asset tracking took off as an RFID application, gaining market traction in industries such as oil, gas, mining, IT, and nonpharma healthcare.
- RFID was successfully used in the Florida court system and by the Alameda County, California registrar of voters to simplify and guarantee labor-intensive government procedures.

More than any other year, 2008 proved to draw a line in the sand. "There's a decreased focus on performance," said Joe White, COO of RFID Global Solution, and formerly a VP with Motorola. "[Enterprises have] realized that RFID has become an internal process improvement technology and external value-add for consumers." RFID has a good many variables, including read rate (by how many RFID tags per second) and sheer accuracy (measured in percentages). The rate may need to be very high, for example, to wave a reader past a rack of clothing, or books on a shelf, or high-volume production output. Accuracy should approach 100 percent. In the early Walmart supplier implementations, a supplier might start with a horrifyingly low accuracy of around 60 percent. After much learning, process tweaking, and improvement of the technology, accuracy of 99 percent is commonplace.

As American Apparel's Zander Livingston observed, the RFID market was

> Very quiet. It wasn't dying, but it was in this lull. Four or five years ago, people had great expectations for RFID and a few people kicked the tires, but nobody really took off on it and even Walmart slowed the whole scale of the project. When I started this project in 2007, retailers were looking closer at RFID when news came out about it being used in more industries, in cars, auto parts, even luggage. There were a lot more pilots going on, prices had come down and out-of-the-box solutions became much more practical. So people who looked at it five years ago were looking again. There's energy around it that has a lot to do with our project,

with which American Apparel went public in 2008. Since then, Livingston has been on the road constantly, evangelizing the results.

2008 was also the year in which RFID migrated from being an information technology or IT project to an operational discipline, as we described earlier. "Now the IT guys are saying, 'It works, the tags and readers work; now it's up to you ops guys to leverage that technology in the infrastructure,'" said White, which is a positive step. "When operational folks get involved, they look at the business benefits, the stakeholders, and the data. I would characterize RFID as no longer an IT project, but an operational discipline, which calls for less of a focus on core technology, and more on the business benefit."

At that point, instead of acting as technology for technology's sake, Wireless becomes one of the tools in a kit, which is where White believes it must go. "We see great opportunities for imaging, bar coding, direct part marking and Wi-Fi based tracking, not just RFID." White observed of the RFID World show that "There's no Bar Code World, no Wi-Fi World. The day we don't have an RFID show is when the technology is mainstream and adopted." It will get there.

## Wireless = Modern

Wireless and Web 2.0 are inevitable in any enterprise, and growing exponentially. They are simply the way of doing business. A modern enterprise that disallows cellular phones and does not bother to have a Web page would be both absurd and noncompetitive. These tools are as inseparable from the modern enterprise as Dr. Jekyll is from Mr. Hyde.

We say "Jekyll and Hyde" because Wireless and Web 2.0 represent both tremendous opportunity and liability. The same mobile computing and Web 2.0 that enable a cost-effective viral marketing campaign also open the door for a viral security breach. Also true, the Tactical Wireless tools that are supposed to minimize unnecessary labor can create more labor than they eliminate, between dragged-out implementations and lengthy learning curves. That happens when the tools are adopted in the passive sense, when an enterprise does not plug the tools into its value streams with intent and with clearly defined expectations for results. Companies that achieve success with Wireless know just which problems they need to solve. For example, Hong Kong International Airport, McCarran Airport in Las Vegas, and Heathrow Airport in London all needed to improve baggage tracking, and did so with RFID. The International Air Transport Association (IATA) estimates the accuracy of bar-codes in baggage handling of 80 to 90 percent, versus up to 99 percent for RFID. At those figures, IATA estimates the airline industry could reduce lost or delayed bags between 12 and 15 percent and save $760 million per year with RFID baggage tracking.[17]

Mobile computing devices like laptops and BlackBerries are perceived as productivity tools, but few organizations know just what they expect in return from them. As we show in Chapter 4, "The Democratic Frontiers of Lean Wireless," London's Thames Valley Police Force used a specialty application called OnPatrol on BlackBerry devices to reduce officers' time at police stations by 30 percent, and for instant access to warrants and court order databases.[18] The results are measurable, and justify the costs of the BlackBerry devices, software, and data plans.

Wireless and Web 2.0 utilities are far more economical and standardized than they were in 2000. RFID, for example, may have cost more than the problems it solved in 2004, but if you revisit the value stream map with Wireless in mind, there are huge opportunities. Wireless has reached a maturity and economy of scale such that enterprises are implementing wireless local area networks (WLANs), voice-over-Internet protocols (VoIP), and even voice-over-Wireless networks (VoWANs). Apparel retailers use RFID to maintain inventory with nearly 100 percent accuracy, heavy equipment manufacturers are using GPS to locate stock in their own yards, hospitals use RFID to count and bill for inventory, and filling stations use a transponder on your dashboard to bill you for a fill-up, rather than take your credit card.

Moreover, there is an opportunity to leverage the economies of scale that have arisen from Wireless entering consumer areas. Consumer-grade wireless networking—a Wi-Fi Internet router—is available in homes for as little as $30, and an enterprise-grade Wi-Fi router can cost as little as $200. Similarly, consumer-grade GPS widened its appeal and fostered lower costs (well under $100), and enterprises use it routinely in logistics and in yard management.

Wireless is inevitable, as is Web 2.0, like it or not. Columnist Rob Preston of *Information Week* advised readers not to be "dragged kicking and screaming onto this emerging business technology campus ... in addition to making users happier and more productive these tools can be a lot cheaper, freeing up dollars for other game-changing efforts."[19] Companies as diverse and successful as Coca-Cola, Nike, Motorola, and Procter & Gamble *encourage* employees to use Wireless consumer technologies such as smartphones, and Web 2.0 applications such as social media. With these tools, they expect employees to collaborate, manage projects, and form teams to solve business problems (in essence, iKaizen).[20] Walmart, Sam's Club, the DoD, libraries, and hospitals are implementing Wireless to achieve efficiency and profitability. Partnering with the Lean camp can help to ensure that they achieve those goals.

## The Wired to Wireless Progression

Although the momentum is recent, we have migrated from wired to Wireless steadily since the early 1900s. Radio was an absolutely disruptive technology in its day, if we understand a disruptive technology to have a momentum of its own, reaching farther than it was intended to do and providing benefits beyond those anticipated.

Radio was exciting and fashionable in the early 1900s. Two hotels, one in Manhattan and one in Chathamport, Massachusetts, named themselves the "Radio Hotel." In 1919, Italian immigrant Antonio Pasin renamed his Liberty Coaster, the child's wagon that he produced by hand, the Radio Flyer. To Pasin, the word "radio" implied all things modern, fun, and forward thinking. President Calvin "Silent Cal" Coolidge, on December 6, 1923, delivered the first State of the Union address (before it was called that) over radio, from a studio in New York City. By all accounts he trusted neither radio nor Congress, and his oration was somewhat pained.

The same broadcast-tower technology that sent Coolidge, *Amos 'n Andy*, and *Baby Snooks* into homes later transmitted *I Love Lucy*, Edward R. Murrow, and *The Honeymooners* to televisions. Radio comic Fred Allen in 1947 was unimpressed; he dismissed TV as "talking furniture" on which "people who haven't anything to do watch people who can't do anything."[21] Within two years of that statement, he would lose his radio program when his listeners migrated to television. The wires were not completely gone; in 1948, cable television reached remote areas where broadcast towers were impractical. Still, the wires thinned down or disappeared completely as television migrated from coaxial cables to fiber-optic cables to satellite transmission.

All the while, telephones went from wired to wireless. If you had one, and most Americans in the 1930s did not, your telephone began as a wall-mounted, hand-cranked, hardwired party line that you shared with everyone on your street. These evolved to private lines; then, in the 1980s, to four-pound brick-sized "portable phones" that you carried in a satchel; and finally to all-in-one cellular phones and smartphones with instant messaging and Internet access that we carry in our pockets. As of late 2006, roughly 77 percent of the world's population lived within cellular telephone range, and a 2008-model cellular phone has about the same computing power as a desktop computer from 2000. 2005 was a watershed for Wireless, in which laptop computers with built-in Wireless cards outsold desktop computers.[22]

Now, let's look at the evolution of product and inventory tracking. Figure 1.4 represents the progression of tracking methods, from notebooks and clipboards (completely hands-on) to bar coding (contactless, but still requiring an operator), to RFID, RTLS, and GPS; these are absolutely contactless and hands-free. When implemented well, they require no labor whatsoever. In time, anything that can be Wireless, from a telephone to a light switch to a routine task, will be Wireless.

## Wireless, Near and Far

To understand how Wireless the world already is, consider this continuum of technologies in terms of sheer distance:

- Bluetooth and ultrawideband (UWB) personal area networks (PANs) connect us to devices within three feet.
- 802.11 Wi-Fi and WLANs connect us up to several hundred feet (inside a building or a Starbucks).

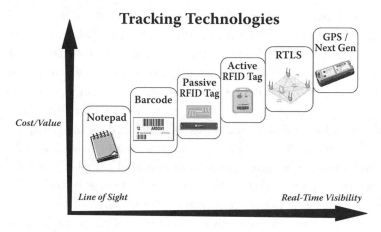

**FIGURE 1.4**

The continuum of tracking technologies, from pad-and-pen through global positioning. (Courtesy Rush Tracking Systems)

- WiMAX, the "metro area network" or MAN, connects us up to a range of (reportedly) 30 miles.
- Cellular phones connect us with no practical limit on range, using a mesh network of towers.

These Wireless technologies vary by frequency and power. PANs, for example, use frequencies capable of transferring large amounts of data but with low power, so the data is not sent to people miles away (and no one can listen in on your phone call when you use a Bluetooth headset). A headset or printer can get by with a PAN; a college campus, city, or military base requires WiMAX.

What the technologies increasingly have in common is their use of the Internet as a method of information exchange. The Internet connects the enterprise and its customers in a way that is simply impossible otherwise; it provides two-way immediacy. Consider the case of American Apparel (or any other retailer/e-tailer):

- A consumer can place an order immediately, and at any hour of the day.
- The e-tailer monetizes the purchase immediately.
- The e-tailer can create promotions, adjust prices, and add or subtract inventory immediately.

Microsoft's Bill Gates called it correctly: "The Internet will help achieve 'friction free capitalism' by putting buyer and seller in direct contact and providing more information to both about each other."

## The Retail Driver

Although American Apparel demonstrated the benefits of an RFID-enabled inventory replenishment system, its customer base is very focused; it lacks the scale to push consumers to say, "I'm changing the way I buy."

Walmart, however, has the scale, the power, and now, the technology. In 2004 Walmart famously mandated that its suppliers put RFID labels on incoming shipments, using ultrahigh-frequency (UHF) RFID with tags written using the EPCglobal Generation 2 or ISO 18000-6C standards. These were the standards hammered out by technology providers (including Texas Instruments, Intermec, Motorola, etc.), by users (including Walmart, Tesco, Johnson & Johnson, etc.), and by GS1, an international not-for-profit association dedicated to global standards for supply-chain solutions. Walmart uses them to track inventory, avoid stock-outs, manage supplier relationships, and the like. As of October 2008, Walmart announced the status of EPC/RFID adoption as follows:

- Every Sam's Club store (all Walmart owned) in the United States was enabled for EPC/RFID.
- More than 3,000 Walmart stores were also EPC/RFID-enabled.
- All uniform pallets going to Sam's Club distribution centers in Kansas City, Missouri; Dayton, Texas; Searcy, Arkansas; and Villa Rica, Georgia must have EPC/RFID tags.
- All uniform and mixed pallets going to the Desoto, Texas distribution center must have EPC/RFID tags.

For over three years, Walmart urged suppliers to adopt EPC/RFID technology, at first voluntarily and then by charging a tag application fee. That is, unless suppliers tagged the items as required, they would be charged by Walmart for the tagging (at $2 per pallet load, in the case of its Sam's Club suppliers). Many suppliers saw the benefits of the technology and began adoption. The most forward-thinking companies actually revised their business processes and achieved significant benefits using EPC/RFID.

Others did the minimum by affixing tags at the last possible moment before shipping to Walmart or Sam's Club, a practice nicknamed "slap and ship" by RFID integrators. The reward of slap and ship is compliance with Walmart and a continuing relationship. The suppliers incurred expense, without any process improvement or return on investment. Regardless of the reason, suppliers simply did not all adopt the technology at the same pace, and Walmart grew impatient.

In February 2009, Sam's Club started placing EPC/RFID tags on selling-unit items that were not tagged by the manufacturer. This removed the final barrier to a fully EPC/RFID-enabled infrastructure. At that point, the retailer's next step was fairly obvious: RFID-enabled point-of-sale, the "holy grail" for retail. When Sam's Club stores will deploy EPC/RFID-enabled point-of-sale systems and eliminate checkout lines, it will change consumers' expectations so much that it will alter their buying habits. Checkout lines will be, at least in theory, a thing of the past. It only takes one holiday shopping season for consumers to spread the word, "Just push your cart through the checkout, swipe your debit card, and you're done." (As of this writing, this convenience has yet to be realized at Sam's Club stores.)

Long lines were the biggest barrier to the growth of club stores and in a recessed economy, consumer loyalty is to their wallets. Retailers that have, in the past, chosen not to adopt EPC/RFID cannot adopt the technology fast enough to compete. They not only lack the infrastructure, but they lack the tagged products. Already staggering from a difficult economy, dozens of major retailers are being forced into bankruptcy or to downsize, and Walmart captures the majority of the market share.

Because the Sam's Club distribution centers are also EPC/RFID-enabled, their product throughput is three to six times faster than their bar-code-based competitors'. Not only can they get products to the store more rapidly, but a single distribution center can support more retail locations. Their expansion is only limited by the rate at which they can build stores or move into the competitor's vacated facilities.

The retail giant's approach has also forced more retailers to adopt a "club" approach to business, wherein consumers must accept the terms and conditions of membership prior to being able to make a purchase. This allows the retailer to eliminate the privacy issues and any legal risks associated with RFID, which are largely imaginary. The perception cultivated by anti-RFID advocates is that a malefactor with an RFID reader can point it at your pocketbook or pockets and identify anything in there including medications, and read your RFID-enabled credit card, passports, and so on. From this reading, the selfsame malefactor has your address, social security number, and so on, which is impossible; there is no more information on those cards than a number. Without the databases in which those numbers sit, the numbers are useless. Our malefactor would need to read your pockets and steal several databases, which is theoretically possible but has never happened.

Still, all the legal protection the retailer requires is granted when consumers sign the terms of membership. The choice to a club member is simple: accept EPC/RFID or don't shop here. Many consumers willingly adopt EPC/RFID and there is absolutely nothing the privacy advocates can do about it.

The retailer's growth has catastrophic implications to some manufacturers as well. With the increase in demand, Walmart expects faster product replenishment from suppliers. Those suppliers that have adopted EPC/RFID can increase their production and handle the speed at which products are ordered. Manufacturers that did not adopt EPC/RFID are unable to fulfill their orders in a timely fashion. To prevent out-of-stocks, Sam's Club turns to other suppliers with like products. Now that consumer loyalty has shifted to the club store with (in theory) no lines, smaller manufacturers that adopted EPC/RFID take up the slack and begin capturing market share from their larger, less-agile competitors, once again, in theory. When this comes to pass (and it is well on its way) the retail landscape will be changed forever.

## Wireless Growth Is a Step-Change Evolution

Time will tell, but the advancement of Wireless and Web 2.0 has all the makings of an economic and technological step change on a par with immunology and automotives. Wireless technology may seem routine to us today, but let's put it in the timeline of history so that we can fully understand the impact.

There was, in essence, little economic and technological growth from the first century AD until the Middle Ages, around 1500 AD.[23] The changes over 1,500 years in standard of living, economic growth, and technology were practically indistinguishable; someone plucked from the year 1 AD and plunked into the year 1500 AD would see little that she would not recognize or could not understand. Foot soldiers in the armies of Augustus Caesar in 1 AD and Richard II in 1500 AD all fought with swords.

Now consider the exponential differences in class, economy, and technology, before and after the 1500s. Between 1 AD and 1500 AD, there was perhaps one-half of one percent rate of economic growth in one generation, whereas today, we see growth (or decline) of that scale in one year. Barry Asmus, senior economist at the National Center for Policy Analysis, cites such factors as:

- The printing press
- The concepts of savings, investment, and private property
- The rise of physics, chemistry, and biology
- The rise of the United States and its free-market and trade influence upon India, China, and Latin America, which are now emulating U.S. practices of privatization, lower taxation, lower tariffs, and free trade[24]

Consider the democratic nature of these advances, which will also come to bear in Wireless momentum. Before the printing press in the late 1400s, only a king or bishop owned a book, which was usually the Bible, and he would wait a decade or more for it to be handwritten by monks. By 1722, even the downtrodden owned copies of *Moll Flanders*, thrilling as its raunchy heroine defied both aristocracy and the gallows. Printing was a disruptive technology, growing exponentially and becoming democratic when profit could be made from the lower classes.

For millennia, savings, investment, and private property were for the gentry. Not until the Pilgrims in America did humble folk own land or turn profits for themselves; now almost every class of American has some investment, even if only in an IRA or 401(k) account. These concepts were disruptive for their far-reaching impacts. The concepts of investment and private ownership have spread worldwide. Incidental to this is the supposed "deterioration" of the family unit, likely because individuals can be self-supporting and the elderly are not as dependent upon the young as they once were. These were tremendous disruptions to the order of the world, and ones that occurred in less than 300 years.

Let us turn now to work and computing, which both Wireless and Web 2.0 touch. The productivity of the U.S. labor force alone doubled between 1993 and 2008, as opposed to growing less than two percent between 1900 and 1995.[25] Likely, this had much to do with advances in both business practices and technology. The 1980s and 1990s saw continuous improvement becoming the norm, business standardizing its platforms and interfaces (on Microsoft Windows), and of course, the rise of the Internet as both a business and consumer utility.

And rise it did, and continues to do. A 2008 IBM Institute for Business Value study revealed that the rate of online and mobile content doubled from 2007 to 2008.[26] Among 2,800 survey respondents in six countries, a full 60 percent used social networking such as Facebook and MySpace, and 40 percent used Internet data plans in their mobile services (chiefly on cell phones and PDAs).

Wireless momentum is not linear, gaining one user at a time; rather, it is exponential. And unlike computing, which was originally driven by government and industry, Wireless has the force of consumer demand and use behind it.

## Brace Yourself: Wireless Momentum Will Only Quicken

Fred Allen was not a fool for dismissing television: he simply underestimated its disruptive significance. IBM Chairman Thomas Watson made a similar miscalculation in 1943, when he famously said, "I think there's a world market for maybe five computers." And in 1977, Digital Equipment Corporation President Ken Olsen said that "There's no reason anyone would want a computer in their home."

Television and home computing defied market forecasts, and Wireless and Web 2.0 are poised to do the same.[27] Like computing itself, Wireless and Web 2.0 are governed by Moore's law, which Intel co-founder Gordon Moore postulated in 1965. Moore predicted that the number of transistors on a chip would double about every two years.[28] In the mid-1960s, a 512 K computer filled a room, and only Apollo spacecraft had onboard computers. Now computers are measured in gigabytes and terabytes of processing power, and as of 2008, all automobiles sold within the United States are required to use an onboard diagnostic computer (by ISO 15765-4 standards). Moore's law makes computing cheaper and more accessible, thus, more democratized. But even the forward-thinking Moore underestimated the speed of this progression.

In each of those miscalculations (from Fred Allen, IBM, DEC, and even Moore), someone failed to predict the democratic effect. DEC's Ken Olsen made his statement when a financial software program involved an 18-month outsourced implementation; he couldn't have conceived of Quicken Home Edition on a CD-ROM. But once computing technology was more efficient and less expensive, more people demanded the technology, and they used it in ways no one predicted, such as gaming. The monolithic UNIVAC I computer accurately predicted the outcome of the Stevenson/Eisenhower presidential race, but held little promise of fun and frolics.

That same IBM Institute for Business Value study observed something very startling: the decline of television as the primary media device.[29] "This year's study [2008] found large scale adoption and usage of digital content services accessed via the PC and mobile phone." And why not? Anyone eager to see an episode of *The Office* may now watch it on a Wireless mobile device, anywhere and at any time. True, they might have to watch it on a tiny screen, but it is immediately available, the commercial breaks are less disruptive, and total viewing time is 22 minutes versus the full half hour. Similarly, they will download music in lower-quality compressed formats but in quantity, rather than spend hundreds of dollars on CDs.

Expediency does not supplant quality: expediency *is* quality, to Web- and Wireless-savvy people. The Institute observed that, although digital content services adoption is widespread, interactivity through features such as user ratings tools and video uploads is "primarily concentrated among the more digital savvy consumers."[30]

Put it all together, and Wireless changes the daily lives of consumers. Marvelous. But what has that got to do with business?

## SURPRISE—THE MODERN NETWORK IS ALREADY HERE—WE NEED ONLY PLUG IN

If you say, "We need a real-time location system" or, "We need a wireless sensor network" to a modern operations officer, she will in all likelihood groan. Say instead, "We could use an 802.11 Wi-Fi system, you know, the kind the college kids use at coffee shops," and that seems a shorter and less expensive leap.

The Wireless technology vendors of, for example, real-time location systems and Wireless sensor networks recognize that, and are plugging into Wi-Fi and other Wireless systems like it. The result is that Wi-Fi or another similar system becomes, if not the enterprise backbone, then its nerve path, and every Wireless device, from a laptop to a PDA to an RFID reader, plugs into it.

## How Lean Is This?

Consider two configurations of an automotive plant on the following pages, the first being typical, and the second, theoretical. The first (Figure 1.5) incorporates all the modern Wired and Wireless technologies, in a jumbled mess of "nerve paths" that cross but do not intersect. Each device must be integrated with the enterprise system. Having more Wireless nodes requires better connectivity and integration, else, the information will simply go to waste. Figure 1.6 shows the same technologies attached to a single Wi-Fi nerve path. This single system in turn integrates with the enterprise system, perhaps an SAP or Oracle system.

To understand just how cost effective this is, the U.S. Department of Energy estimates that "installation of wiring can represent 20 percent to 80 percent of the cost of a sensor point." Installation cost is typically calculated as follows:

*Cost of the wire* (\$ per linear foot or meter) + *Labor time for installation* (measured in linear foot or hourly) + *Misc. costs* (changes to facility, securing wires, ensuring compliance with safety standards)

Depending on the location of the sensor, installation can cost between \$200 and \$400 per sensor, plus wiring cost, which can be \$200 per foot depending on the type of cable. In addition, the wiring can be expensive to service and maintain.

If the enterprise system is a "backbone," as it is commonly described, the Wireless network is a sort of "nerve path" for virtually any device, system, or process, including:

- Telephony, using VoIP
- Sensors
- RTLS

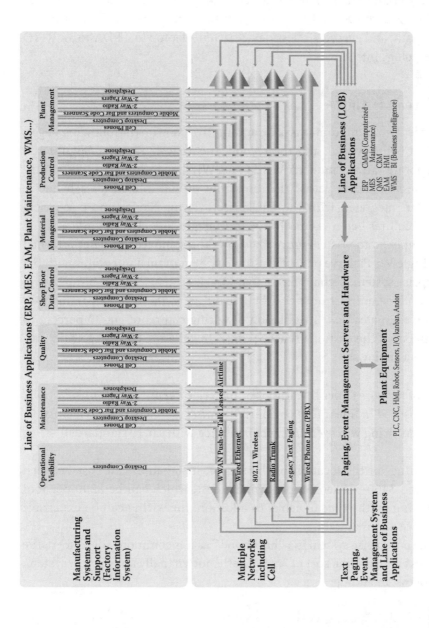

**FIGURE 1.5**

Today's typical automotive plant has modern Wired and Wireless technologies, but no central nervous system. (Courtesy Motorola, Inc.)

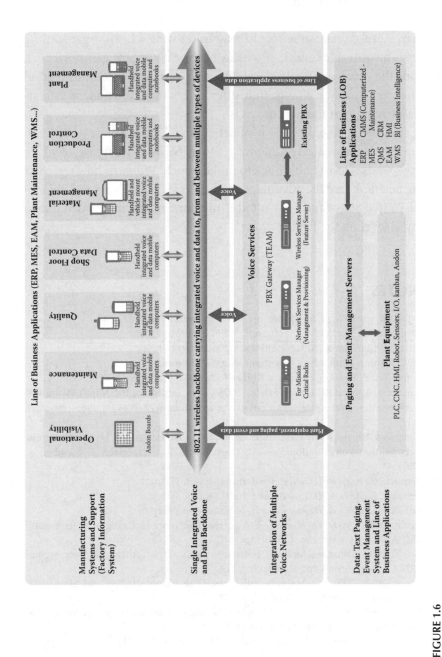

**FIGURE 1.6**

A single readily available nervous system, using readily available devices. (Courtesy Motorola, Inc.)

- Mobile computers and PDAs
- Line-of-business applications, such as computer maintenance
- Event management and messaging

Clearly, Wireless has some role to play in creating a Lean organization. It achieves measurable results, as we saw in the case of McCarran Airport, it extends the reach of human decisions, and it has a momentum behind it, unlike any other step change in history. Every industry, and every consumer, is driving that momentum.

And, frankly, business needs it—particularly U.S. business—if it is to compete on a global scale. In Chapter 2, "Why Now?" we explore why U.S. industry spends almost half a buck to make a buck. We show how the costs of doing business give the advantage to nations such as China and Taiwan over the United States and the United Kingdom, and the role that Wireless can play in leveling the field.

## ENDNOTES

1. Tagline at www.americanapparelstore.com (accessed January 31, 2009).
2. Retail Fashion Market RFID Solutions. ©2007, Motorola, Inc.
3. Retail Fashion Market RFID Solutions. ©2007, Motorola, Inc.
4. Retail Fashion Market RFID Solutions. ©2007, Motorola, Inc.
5. Livingston, Z. Panel discussion at RFID World 2008, Las Vegas, NV.
6. "American Apparel RFID Case Study." 2008, Avery Dennison, Inc. http://www.ibmd. averydennison.com/solutions/american-apparel-rfid.asp (accessed September 1, 2009).
7. "American Apparel RFID Case Study." 2008, Avery Dennison, Inc. http://www.ibmd. averydennison.com/solutions/american-apparel-rfid.asp (accessed September 1, 2009).
8. Heinrich, C. 2005. *RFID and Beyond.* Indianapolis: Wiley.
9. Heinrich, C. 2005. *RFID and Beyond.* Indianapolis: Wiley.
10. Heinrich, C. 2005. *RFID and Beyond.* Indianapolis: Wiley.
11. Rio, R. 2008. "Wireless Mobility Enhances, Lean Manufacturing and Six Sigma." ARC Advisory Group, Inc., Cambridge, MA.
12. Rio, R. 2008. "Wireless Mobility Enhances, Lean Manufacturing and Six Sigma." ARC Advisory Group, Inc, Cambridge, MA.
13. Rio, R. 2008. "Wireless Mobility Enhances, Lean Manufacturing and Six Sigma." ARC Advisory Group, Inc, Cambridge, MA.
14. Rio, R. 2008. "Wireless Mobility Enhances, Lean Manufacturing and Six Sigma." ARC Advisory Group, Inc, Cambridge, MA.
15. Taub, S. 2004. "Deep six for Six Sigma?" *CFO* Magazine, May 18.

16. ABI Research, Inc. 2008. "Global RFID Market to Reach $5.3 Billion This Year." http://www.abiresearch.com/press/1284-Global+RFID+Market+to+Reach+$5.3+ Billion+This+Year
17. http://www.rfidupdate.com/articles/index.php?id=1544
18. Research in Motion. "Cape Breton Police Force Leads by Example with a BlackBerry Solution: Case Study." Xwave.com, 2007, http://www.xwave.com/files/credentials/ CapeBreton_lob_final.pdf (09 January 2009).
19. Preston, R. 2008. Down to business: Consumer technology and the IT democracy. *Information Week*, November 17.
20. Preston, R. 2008. Down to business: Consumer technology and the IT democracy. *Information Week*, November 17.
21. Allen, F. 1997. In *Encyclopedia of Television*, ed. H. Newcomb, pp. 63–65. New York: Fitzroy Dearborn.
22. Hiemstra, Glen, 2005. Outlook 2006. http://www.futurist.com/articles/future-trends/ outlook-2006/ (accessed January 9, 2009).
23. Asmus, Barry. 2008. Podcast: America's Six Unstoppable Trends. http://knowledge. wpcarey.asu.edu/article.cfm?articleid=1695 (accessed January 9, 2009).
24. Asmus, Barry. 2008. Podcast: America's Six Unstoppable Trends. http://knowledge. wpcarey.asu.edu/article.cfm?articleid=1695 (accessed January 9, 2009).
25. Asmus, Barry. 2008. Podcast: America's Six Unstoppable Trends. http://knowledge. wpcarey.asu.edu/article.cfm?articleid=1695 (accessed January 9, 2009).
26. Mandese, J. 2008. IBM Study Finds Adoption of Online, Mobile Content Doubling. http://www.mediapost.com/publications/?fa=Articles.san&s=94890&Nid=49437&p =426639 (accessed January 9, 2009).
27. Knowledge@Wharton. 2008. Why an Economic Crisis Could Be the Right Time for Companies to Engage in "Disruptive Innovation." http://knowledge.wharton.upenn. edu/article.cfm?articleid=2086 (accessed January 9, 2009).
28. Asmus, Barry. 2008. Podcast: America's Six Unstoppable Trends. http://knowledge. wpcarey.asu.edu/article.cfm?articleid=1695 (accessed January 9, 2009).
29. Mandese, J. 2008. IBM Study Finds Adoption of Online, Mobile Content Doubling. http://www.mediapost.com/publications/?fa=Articles.san&s=94890&Nid=49437&p =426639 (accessed January 9, 2009).
30. Mandese, J. 2008. IBM Study Finds Adoption of Online, Mobile Content Doubling. http://www.mediapost.com/publications/?fa=Articles.san&s=94890&Nid=49437 &p=426639 (accessed January 9, 2009).

# 2

## Why Now? Lean Wireless and the Costs of Doing Business

Global companies labor under the "costs of doing business," but Lean Wireless can help recoup those costs. Some of those costs, such as taxation, are fixed and beyond a company's control. But a company can control certain variable costs such as labor, IT infrastructures, and repetitive processes.

Both Wireless technology and the continuous improvement discipline target variable costs; together, they provide a methodology of creating and applying business rules that drive down variable costs.

### THE UNBEARABLE COST OF BUSINESS

If it costs an American business $.48 to produce a dollar's worth of goods (we show how, shortly), then in theory, that business profits by $.52 per dollar, which we know very well is untrue. A business is lucky if it does not spend *more* than a dollar to produce a dollar's worth of goods.

That $.48 represents only the costs a company can't control, such as taxes; the other $.52 goes to variable costs over which it does have some control, such as inventory and labor. Both Lean practices and Wireless technology target that $.52; but as we've seen, the camps don't cooperate at all, and barely speak each other's language.

Lean is a proven methodology, Wireless an increasingly proven bag of tools. The two combined offer a uniquely strong value proposition. It is our belief that now—in 2010, and in the throes of a recession/depression/

downturn—is the time to fuse Lean and Wireless, and for the Lean and Wireless camps to actively cooperate.

---

## THE IMMOVABLE STRUCTURAL COSTS

That $.48 figure is called the raw-cost index, or RCI,* and comes from an exhaustive 2008 study by the Manufacturers Alliance (MAPI) and the Manufacturing Institute, which is funded by the National Association of Manufacturers. Manufacturing is, of course, not the only sector in business, and structural costs are common to all sectors. MAPI defines structural costs as "those out of manufacturers' direct control," which include:

- Corporate tax burden
- Employee benefits
- Tort litigation
- Regulatory compliance
- Energy costs

The U.S. government could make it easier on U.S. businesses. We'd save a lot of money and time to market without FDA approval for pharmaceuticals, or Sarbanes–Oxley requirements for financial disclosure, or HIPAA requirements for patient confidentiality. But this is the United States: we're used to high standards, and neither our legislature nor private sector is calling to rescind these protections.

And the burden could be worse, much worse. The study compared the United States to nine industrialized countries: Canada, Mexico, Japan, China, Germany, United Kingdom, South Korea, Taiwan, and France. It costs a U.K. company about $.71 to produce $1 worth of goods to our $.48. Conversely, a Chinese company spends only about $.20.

These figures have been far worse for the United States. Recently these costs comprised a staggering 31.7 percent disadvantage to U.S. business in 2006; thus the structural costs to a U.S. business were 31.7 percent higher than the average among our trading partners; when they spent $1 million,

---

* The cost to produce $1 of goods.

a U.S. company spent $1.37 million. That disadvantage dropped to 17.6 percent in late 2008. MAPI and the Institute cite three factors for the decrease:

- High investments and productivity in the United States
- Policy changes in the United States, such as tort reform, and leveling out of healthcare insurance premiums
- Rising costs elsewhere in the world, which produce a leveling effect

Observe that the only direct influence that a company has is on the first factor: through harnessing new technology, companies can raise productivity, innovate new products and processes, and manage high structural costs. In fact, ABI Research found that although U.S. industry as a whole was cutting costs, the average RFID system expenditure among survey respondents in 2009 increased 28 percent over 2008, from $1.7 million to $2.2 million. Furthermore, growth in RFID is not limited to supply chain management and industry, but also spans healthcare, commercial services (e.g., legal, lodging, and education), and government and utilities, among other verticals.[1] (See Figure 2.1.)

## The Tax Burden

Corporate taxes are the most monumental and immobile of structural costs, and a particularly heavy brick in a U.S. company's load. The tax burden on a U.S. company is about 40 percent, and has remained so for years. Of those nine trading partners, only Japan tops the United States, at 40.7 percent. China and Taiwan are taxed at 25 percent. And the study takes into account only nine industrialized trading partners, whereas the U.S. tax rate is nearly four times that of Ireland (at 11 percent), for example, which is emerging as a software brain trust. This, opines the National Association of Manufacturers, "cuts into the competitiveness of American businesses that operate in a global market."[2]

What happens elsewhere in the world is also out of a company's control, but it is benefiting U.S. companies. Other industrial countries and emerging economies are seeing sizable wage increases; hourly pay in Mexico, for instance, has leaped 55 percent in seven years. Both Canada and the United Kingdom have seen higher healthcare benefit costs, as employees require private insurance to supplement fair-to-middling national healthcare.

**Table 5-1**

**Primary RFID System Adoption Drivers**

**End-User Respondent Average Ratings by RFID History/Usage, Excluding Non-Users and Non-Evaluators (1 = Not Important; 5 = Extremely Important)**

| Category | All Respondents (Excluding Non-Users and Non-Evaluators) n = 60 | System Installed n = 18 | System Installed and Piloting/ Evaluating Additional Applications n = 19 | Currently Piloting, Testing, or Evaluating n = 23 |
|---|---|---|---|---|
| Business Process Improvement | 4.23 | 4.22 | 4.42 | 4.09 |
| Ease of Scalability/System Extendibility | 4.15 | 3.94 | 4.37 | 4.13 |
| Ease of Integration | 4.07 | 3.67 | 4.32 | 4.17 |
| Removal of Human Intervention | 4.02 | 4.11 | 3.95 | 4.00 |
| Performance in Harsh Environments | 3.77 | 3.50 | 4.00 | 3.78 |
| Technology Price Declines | 3.73 | 3.22 | 4.11 | 3.83 |
| Introduction of Business Intelligence & Data Analytics Tools | 3.72 | 3.50 | 3.95 | 3.70 |
| Standards Development/Advancement | 3.63 | 3.44 | 3.84 | 3.61 |
| Customer and/or Regulatory Compliance/RFID is Mandated | 3.28 | 3.28 | 3.79 | 2.87 |

*Source:*   ABI Research

**FIGURE 2.1**

RFID end user survey: end users segmented by primary vertical market. (Courtesy ABI Research)

Tort costs are rising in some European countries, and the United States has passed tort reforms to limit liabilities.

China's structural costs have risen considerably in the last three years (that $.20 used to be $.13), chiefly due to new pollution control costs, but China still enjoys a considerable advantage. This upper hand has helped them to surpass Canada as the largest supplier of imported goods to the United States ($321 billion versus $313 billion from Canada in 2007); and since 2000, China's share of total U.S. imports has doubled (from 8.2 percent to 16.5 percent).

The short story is that the world is, in essence, "going American," which is helping the United States. But the playing field isn't yet level, and that 17.6 percent disadvantage is costing the United States in competitiveness. United States companies cannot be productive or profitable enough.

## A MORE INCLUSIVE PICTURE—VARIABLE COSTS

If $.48 is the RCI in the United States, then a U.S. company should profit $.52 for every dollar spent, but of course there are far more costs of doing business than those structural costs. A short list includes:

- Inventory
- Labor
- IT infrastructure
- Governance, risk, and compliance (GRC)
- Production (in-house or outsourced)
- Brick-and-mortar offices (own or lease)
- A Web presence
- Sales and marketing
- Transportation
- Human resource management (in-house or outsourced)
- Logistics
- Raw materials and services
- Security

With all those costs, most companies will have very little of that $.52 left over (if any). However, although these costs are unavoidable, companies can influence them. In the following sections, we talk about how Lean, Wireless, and Lean Wireless can affect these factors.

## THE LEAN, WIRELESS, AND LEAN WIRELESS VALUE PROPOSITIONS

Continuous improvement targets every one of those variable costs. Continuous improvement, and Lean in particular, looks for less waste and higher efficiency in every corner of a business. In one of the more remarkable Lean stories in recent years, the United States Air Force (USAF) applied Lean practices to its logistics operations. As we described in the Introduction, the USAF eliminated waste and redundancy in its IT applications and freed up more than $347 million to upgrade mission-critical applications.[3]

All the while, technology and automation have targeted those same costs. Recall that MAPI credited U.S. companies with harnessing new technology and raising their own productivity to improve products and processes. United States companies themselves agree, crediting technology and automation for increased productivity. More than 185 organizations from across the world responded to ABI's 2008 survey of RFID usage, sharing their plans, adoption drivers, value propositions sought, and expectations of return on investment (ROI).

The drivers to adoption are a good indicator of their priorities. Among those companies, the four most important drivers were, on a scale of 1 (lowest) to 5 (highest):

- Business process improvement (4.23)
- Ease of scalability/system extensibility (4.15)
- Ease of integration (4.07)
- Removal of human intervention (4.02)

The second and third drivers are *usability* drivers: these companies are anxious to use the technology and now find it feasible. The first and fourth (process improvement and removing human intervention) are their top *organizational* priorities, and perfectly align with the Lean and continuous improvement priorities.[4]

The four top benefits and value propositions driving RFID adoption were:

- Inventory visibility (4.12)
- Efficiency gains/business process improvement (4.08)
- Asset visibility (3.87)
- Labor reduction (3.85)

Here again, each benefit is directly traceable to a Lean value proposition. Both Lean and Wireless fulfill the promise of "doing more with less," and achieving peak efficiency.

Finally, these companies have confidence in Wireless technology. Of those respondents with a system installed or who were piloting or evaluating applications, 42.1 percent expected a 100 percent ROI in less than a year, and 73.7 percent in less than 18 months.

## INFORMATION IS LEAN

Earlier we pointed out how information replaces inventory in Lean; the more tightly controlled an inventory is, the less buffer inventory is required. But, it replaces a good deal more in Wireless practices. Among hundreds of other tasks and objects, it replaces cash transactions, physical inventory counts, wired infrastructures and telephony, and security checks. Wireless technology providers offer new methods of continuous improvement.

Let us for a moment look at inventory. The ideal in Lean is to move a finished product from the production line to the loading dock, bypassing the distribution center, and ultimately, to repurpose the distribution center for more production. A corresponding ideal of Wireless is to automate production replenishment, eliminating both buffer inventory and shop floor inventory, and wasted movement in replenishment; a production cell automatically "phones in" an order when it's needed. The combined value proposition of Lean Wireless is pull-from-demand inventory on the production line, little inventory of finished product, and no more movement than is required to fulfill a customer's order.

This Lean Wireless practice is "eKanban," and it is old news. You will find it at Boeing, Motorola, and Toyota, among others. But electronic work-in-process ("eWIP"), remote process improvement ("iKaizen"), electronic mistake-proofing ("ePoka-Yoke"), and electronic go-see ("eGenchi Gembutsu") are terms yet to be coined, but they are due. These practices exist in pockets, and are achieving the step changes in efficiency that Lean has always targeted. But because these practices are largely Web-, Internet-, and Intranet-based, the terms iWIP and iKaizen are more accurate.

## AMERICA AND ITS COMPANIES REQUIRE WIRELESS STRENGTH

Interestingly, the National Association of Manufacturers (NAM) has set a number of Wireless priorities for both industry and citizenry, which it sees as interrelated; that according to Marc-Anthony Signorino, NAM's director of technology policy, who observes that U.S. manufacturers are increasingly dependent upon broadband technology for just-in-time and Lean manufacturing.

Supposedly, we are far behind the world in telecoms and Internet connectivity. Said Signorino,

> You see all these stats about how the U.S. is 31st in broadband deployment, and we're usually behind countries like Japan and European countries and Korea. The reason we're so low on those lists is that the country is so freaking big, and the others above us are so tiny, with high population densities in small areas.
>
> Not to be crass, but most countries at the top of the lists were completely leveled in 1945, while in Washington DC most buildings that were built in the late 1800s are still there. When you raze a city to the ground and build it over, you can build it with all new technology, from 1950 on. And it's easy to wire a building and give Internet service to 5,000 people at once because you've jammed 5,000 people into a building in a small area like in Tokyo.
>
> So those numbers have to be taken with a grain of salt. We do a phenomenal job of bringing telecoms and wireless and broadband service to all Americans [and to U.S. business]. But the job isn't done.

NAM is campaigning to reform the Universal Service Fund, which was established to bring telephone service to rural areas. Between landlines and cellular technology, that mission has been accomplished. "What is more helpful is to be as technologically neutral as possible," said Signorino. "Cellular, satellite, WiMax, Wi-Fi—all sorts of methods of getting to the Internet. High-speed connections are very important to farmers. You can't buy a tractor or a combine without it." Farmers buy GPS-enabled tractors and dash-mounted combines that tap into servers, which are essential for food traceability. Wireless technology allows the exact location, date, and time of harvest to be recorded onto an RFID tag, which is affixed to the trailer containing the crop. That information moves with the crop to the processing facility and is associated with the final food product we purchase.

Recall the November 2006 *E. coli* outbreak from spinach that sickened people in 21 states. It was extremely difficult for the FDA to trace the origin of the outbreak; now, using Wireless technology, we can know where something was harvested within 12 feet. Clearly, Wireless efficiency and traceability stretches far beyond manufacturing and durable goods.

What manufacturers increasingly require are dedicated frequencies, said Signorino. "One thing we have worked on over the last year, and were successful in, is not so much a telephone issue or data issue, but private radio," the 900 MHz frequency for high-end walkie-talkies for use on production floors, loading docks, and man-down systems. "The people at Boeing have the largest work space on the planet," said Signorino, "the hangars in which they build aircraft, and a problem they have is in locating employees who might be crawling around inside of an aircraft. So they need a radio system that operates with hearing protection, that's integrated with radio in case someone is welded into a plane."

NAM's fear was that frequencies in the spectrum around 900 MHz were coming up for auction, and that a company such as Sprint or Nextel would then license it for commercial service. "We were successful in making sure it was free for unlicensed use. Boeing has a private radio station just to handle these. And you'll find people working in electrical grids use these, and trains use private radio systems to communicate with trains on their lines, wherever they are. We were successful in making sure the bandwidth 900 MHz is left open."

NAM views Wireless as a business enabler, and one that must be readily available to U.S. industry, rather than adding another cost of doing business.

## AMERICAN COMPANIES REQUIRE NEAR-TERM RESULTS (NOT SOLUTIONS)

All of the solution providers who cooperated in *Thin Air* have observed that solutions budgets have tightened. "That is driving them more than ever to buy results," said Vice President of Acsis, Andre Pino, "than wanting to buy technology. They're looking to improve efficiencies and optimize what they've got, because they're all being pushed to pull cost out of the system."

Hence, in 2009, a crop of fast-track offerings aimed at near-term results emerged. For example, in January 2009 Acsis introduced its Optimization Platform, aimed at reducing costs and labor requirements through visibility and control; and its Efficiency Assessment Program aimed at identifying fast-track improvement opportunities in manufacturing and supply chain. The one-day on-site session targets near-term benefits and ROI, and measurable improvement in productivity, data accuracy, and resource utilization. Pino notes that Acsis customers typically expect a six-month ROI.

Similarly, Oliver Wight, the business improvement organization with offices throughout the United States and Europe, introduced a series of seven "FAST TRACK" programs in March 2009. The programs target working capital reduction, order-to-cash time compression, and supplier cost improvement among other goals. The aim, said Les Brookes, the CEO of Oliver Wight EAME, is to "give companies some substantial 'quick wins' with rapid improvements in business performance and a return on investment within months." There is a real urgency in the business environment, and little tolerance for risk taking or experimentation. Wireless technology providers—in fact, all business technology providers—understand that. As a result, you can implement Wireless at a lower cost, lower risk, and higher ROI than you could just five years ago.

### Wireless Provides Fast-Track ROI

Wireless technology providers typically promise 100 percent return on investment or ROI in under six months, and certainly within the first year. At this point in history, it must.

Ekahau, Inc. provides Wi-Fi-based real-time location systems (RTLS), remote sensing, and site visibility solutions. One customer is a nuclear power company, which shuts down once yearly for 35 days for mainte nance. This is a massive undertaking, involving shutting down the reactor; changing pipes, fittings, and valves; and disposing of reactor waste, among thousands of other tasks. Although 35 days is a long time, it is still a scramble.

"Every day a nuclear reactor is shut down in maintenance is a million dollars in forgone revenue," said Ekahau's VP of Business Development Tuomo Rutanen. "Use RTLS to get the right people in the right places at the right time, and you're saving $50,000 an hour, give or take." Ekahau conducted a site survey that reduced shutdown time from 35 to 26 days, and counting. "If they can reduce shutdown to 24 days, they're making $2 million, from a system that might have cost a hundred thousand."

As with continuous improvement, Wireless solutions typically enable other opportunities beyond their original purpose. The power company Rutanen described uses RTLS in its yearly shutdown, but is now imple- menting it in routine operations throughout the year, such as safety and security.

Recall in Chapter 1 that American Apparel had such success with RFID control of inventory, that it is branching out into loss prevention. Wireless success and opportunities beget more success and opportunities, once a Wireless infrastructure is up and running well.

---

## AUTOMATION MEANS MORE WORK—NOT LESS

These are the players, with or without a scorecard: in one corner, a machine; in the other, one Wallace V. Whipple, man. And the game? It happens to be the historical battle between flesh and steel, between the brain of man and the product of man's brain.

**From the 1964 episode of *The Twilight Zone*,
"The Brain Center at Whipple's."**

Despite the fears of the 1950s and 1960s, automation never replaced human workers and eliminated jobs; rather, it enabled and created more jobs. There is not enough labor in the world to do what is already automated. Consider such simple examples of automation as:

- Street lights, which were once turned on and off by technicians in a control room
- E-ZPass, which enables traffic flow at toll plazas
- Automated switching in telephony, which eliminated the task of making physical connections between telephone lines
- Instrumentation and automatic signaling in airplanes, which relay thousands of points of information per minute, looking for exceptions

In each instance, the task was at one time simple enough for a trained individual to handle, but became either too complex or too frequent for an individual to do well, calling for automation.

Still, more people gained jobs than lost them. Smoother traffic flow enabled retailers to open profitable locations in suburbs, and enabled workers to commute beyond their hometowns. Automated telephone switching created the need for service operators and directory assistance; and paved the way for more telephony, and as a result, more operators, engineers, linemen, cell phone engineers, and so on. Automation eliminates tasks, but not jobs. American Apparel, for example, eliminated manual inventory, but still has sales associates.

Wireless, like Lean, does not eliminate work; rather, it eliminates busy work, but generates more value-added, revenue-generating work. But this does not happen by accident; in the worst cases, it does not happen at all. As we show in Chapter 3, Lean experts can smooth the Wired-to-Wireless evolution.

## ENDNOTES

1. Liard, M. and S. Schatt. 2008. *Annual RFID End User Survey.* ABI Research, Inc., New York.
2. Leonard, J. 2008. The Tide Is Turning: An Update on Structural Cost Pressures Facing U.S. Manufacturers. Report by the Manufacturers Alliance/MAPI and the Manufacturing Institute, Washington, D.C.
3. From 2009 interview with Deputy Director Daniel F. Fri, 635th Supply Chain Management Wing.
4. Liard, M. and S. Schatt. 2008. *Annual RFID End User Survey.* ABI Research, Inc., New York.

# 3

## The Lean Wireless Missions

If we accept that the Wireless progression in business is inevitable, then it must be of value to business. It is, but it also creates unique forms of *muda* (the Japanese word for "waste," used in Lean management). Let's call these new forms of waste iMuda.

iMuda is one of several challenges to Wireless that Lean can answer very well, if Lean evolves, incorporates Wireless, and puts it to work. To a degree, Lean has already absorbed Six Sigma, and although Lean traces its roots to manufacturing, it is an increasingly accepted methodology in healthcare, education, retail, and government.

For example, the U.S. Environmental Protection Agency (EPA) adopted Lean practices in 2003, and in May 2009 released its "Lean in Government Starter Kit 2.0." EPA advises state agencies to seek Lean expertise from within the government or from private consultants. The Lean Education Academic Network (LEAN) includes nearly 100 professors from universities around the world; LEAN was formed with the aid of the Lean Enterprise Institute (LEI), and a senior LEI officer sits on the LEAN board.

Thus these fields are ripe for Lean consultants; but those consultants will find that Wireless is already hard at work in those verticals. By 2015, we expect Lean Wireless to emerge as an important subset of continuous improvement, which has boiled down to two simple principles:

- *Eliminate waste.* All that is left is fulfilling, customer-pleasing, revenue-generating process.
- *Automate routine tasks and decisions.* Whatever tasks and decisions are left are those for which a worker's skills are most required.

## ADOPT A NEW PARADIGM:
## RULES BEFORE TOOLS = RESULTS

Earlier in the book we described the continuous improvement paradigm of "rules, not tools," to achieve results. Yet, even the most Lean manufacturers, Toyota included, are not averse to using high-tech tools; nor are enterprises in those Lean frontiers of government, education, healthcare, and retail tool-averse. Simply put, enterprises in all verticals look for results and are neither loyal nor sentimental about how they get them. The paradigm that they practice is "rules before tools = results."

If *kaizen* (the method that involves workers in removing waste and adding standardization to a process) saves money and increases production, an enterprise will use kaizen; if RFID does the same, the enterprise will use RFID. The enterprise cares about the result, and does not care if it achieves the result through a proven methodology or a proven technology. Sometimes too, the technology is all an enterprise needs; it does not require kaizen or DMAIC or consultation. For example, general contractors are as nuts-and-bolts a group as there is, and have the very practical problem of tool inventory, ensuring that the right tools go to a job, then come back from it. Ford and DeWALT joined with ThingMagic to create Tool Link, an inventory system that is an option aboard Ford F-150, F-Series Super Duty pickups, and E-Series vans. The system includes a dashboard inventory of tools, which a contractor has tagged and placed in a DeWALT toolbox, fitted with ThingMagic Mercury5e embedded readers and customized antennas. As Ravi Pappu at ThingMagic describes, the users (the general contractors) are RFID novices, thus, Tool Link must be simple to use and provide immediate results of 100 percent inventory accuracy.

Conversely, technology in some cases is simply overkill, and rules are enough by themselves. "The right technology is not always the latest and greatest," said L. Allen Bennett, President and CEO of Entigral Systems, the real-time location system (RTLS) provider. Bennett was brought into a prison to evaluate whether the prison shop should use a real-time location system to track tools. Although the shop used all manner of tools, Bennett did not recommend Wireless asset tracking. "The shadow box concept worked fine," in which outlines of tools were drawn on a pegboard and signed in and out by inmates working in the shop (presumably with some oversight by guards for tools that could be used as weaponry). "The prison workshop had a finite number of tools in inventory, and the shadow box had been working for fifty years." RTLS would have been overkill in this instance, but was exactly what the prison required to keep track of keys, and it was implemented for that purpose.

Both those examples (tools at a work site or in a prison) are fairly simple ones. They are also closed-loop examples: limited in scope to one organization. Neither rules nor tools by themselves are enough to improve the complex processes of a modern enterprise.

The Manufacturing Enterprise Solutions Association International (MESA) also holds this belief. "Traditional Lean practitioners attempt to say that technology is not needed," said Dr. Ganesh Wadawadigi, vice chairman of MESA and the Senior Director of Operational Excellence Suite Solution Management at SAP Labs. "We are all of the opinion that's not true." By "we," Wadawadigi refers to MESA's board, which is a star chamber of automation giants, including Siemens, Rockwell Automation, and GE Fanuc, plus solution providers such as Invensys and SAP.

MESA believes absolutely in Lean and continuous improvement, but that technology is indispensable in achieving that improvement. Wadawadigi is one of the authors of the *MESA Lean Manufacturing Strategic Initiative Guidebook*. The thrust of the *Guidebook* is the role that technology should play as a company implements sustainable, scalable, Lean business processes. A fallibility of Lean is that it encourages practitioners to set the appropriate size for a process or staff based on current conditions, not future conditions. (In essence, buy the pants that fit you now, not the pants you think will fit you in a month.) Conversely, one measure of a successful technology is its scalability and adaptability. "Obviously you need to look at the processes and make the changes before you adopt the technology," said Wadawadigi, establishing rules before implementing tools, "but technology plays a key role in sustainability and scalability."

MESA encourages what Wadawadigi describes as a holistic view, in which any tool that eliminates waste is valid. In electronics, where defects are measured in parts-per-million and parts-per-billion, Lean and Six Sigma can improve processes, but only technology can make continuous, cost-effective testing and measurement and trending needed to improve those processes. A Six Sigma rule of thumb is that after a point (about Three Sigma), human error introduces more errors than it eliminates.

Thus you require technology to improve a process, else human error in data entry will overwhelm the improvements to the process. (Six Sigma has never been as technology-averse as Lean.)*

Lean consultants generally have a healthy respect for eKanban, but enterprises are fashioning iWIP, iGemba, and iPoka-Yoke; they are continuously improving the methods of continuous improvement, and doing so using technology. And, they consult the technology providers to troubleshoot existing processes, not traditional Lean or Six Sigma consultants. At Motorola, for example, a Motorola subject matter expert will conduct a business process analysis at a customer site along with a Motorola technology partner. "Eighty percent of [those partners] come from the traditional data capture space, and the other 20 percent is a core of smaller RFID-specific companies," said Chris Warner, senior marketing manager, Motorola Enterprise Mobility. "A good breadth of partners has allowed us to approach traditional Lean manufacturing engagements. With a large partner network, we have very specific partners doing very specific applications in specific industries."

Similarly, Ekahau, the Wi-Fi-based RTLS provider, conducts nonintrusive site surveys aimed at finding Wireless opportunities. One field engineer can typically survey 75,000 square feet per day, and two engineers can survey a million-square-foot facility in a little over a week. This illustrates a strong value proposition for Wireless technology over continuous improvement methods: Wireless experts can typically improve processes without disrupting the processes, something that kaizen or a quality circle typically cannot.

If anything, Wireless technology providers differentiate themselves by eliminating any disruptions to business altogether. "People don't always recognize what the cost of disruption may be," said Ekahau's VP of Business Development, Tuomo Rutanen.

> If I'm in a hospital, and I wire in some proprietary RFID system and a sensor in every room, I have to nail it to the wall and put cables on it and string

---

* Moreover, only with technology can a company simulate a process improvement, test its effectiveness, and remove human error. Boeing famously eliminated the need to build a prototype of its 787 Dreamliner by building a three-dimensional virtual Dreamliner and working out the "kinks." Kinks cost real money; rival airplane manufacturer Airbus lost more than $6 billion on its A380 airliner to such problems as cutting wires too short, which caused the wires to snap. Simulation would have prevented that problem before anyone cut a wire. It would also have prevented losing customers to Boeing.

the Internet and power. If I move a tile in the ceiling, the hospital may have to quarantine the room, depending on the law in that state, because you've contaminated the room. People don't always factor in costs like those when they're pricing a system,

but they do rely upon the Wireless providers to recognize the costs and minimize them. This was a common theme among the Wireless providers who contributed to *Thin Air*; they see themselves as providing Lean results, and so do their customers.[1]

"We don't use that Lean terminology," said Vice President and Chief Marketing Officer Andre Pino of Acsis; "We're a process automation company, and we take all of the opportunities for error out of the supply chain by automating those processes," which is of course a Lean mission.

It is clear that continuous improvement along the value stream is hardly an obsolete concept; but companies increasingly trust technology to achieve it. They'll choose technology over methodology—tools before rules.

## Tools Before Rules Is a Mistake

Wireless and Web 2.0 have become noninvasive, easy to use, and plug-and-play, but they are not "technology lite." They require all the forethought, caution, and vetting of any disruptive technology, more so because these technologies are far more rapidly evolving than, for example, automation or telephony.

This is why we propose the new paradigm of "rules before tools = results." This too is why we propose a new business discipline, called Lean Wireless, which combines the strengths of both camps.

If such a discipline arises, what must its missions be?

## Attack Wireless Waste (iMuda)

The rules for Wireless are no different than the rules for any business process. Eliminate all of the waste involved in the process (wasted time, unnecessary movement, pointless repetition, and so on), and whatever is left over is meaningful and value-adding.

Table 3.1 lists the seven common wastes, or *muda*, of lean manufacturing. These seven broad kingdoms of waste apply to any enterprise, not

**TABLE 3.1**

Seven Common Manufacturing Wastes

Manufacturers Can Identify These Seven Wastes to Help Uncover Opportunities for Improvement

1. Overproduction: Producing more than or sooner than is required by an end user or in-process user
2. Overprocessing: Doing more than is required and desired by the customer
3. Motion: Expending excess motion beyond value-added activities
4. Defects: Not only part defects, but any defect in the process
5. Transportation: Unnecessary movement of materials, particularly double and triple handling
6. Inventory: Excess inventory is waste; the only good thing that can happen to inventory is to sell it
7. Waiting: Waiting for information, people, tools, and materials

*Source:* The Lean Learning Center.

just production. The following examples compare a production plant to an insurance company:

- *Transportation*, which applies both to the movement of manufactured goods and to transportation of insurance executives when a phone call or virtual meeting will do.
- *Defects*, which apply to defects in manufactured parts and defects in process. A process defect at an insurance company might be undercharging of premiums, or overpayment on claims, or delayed processing of claims.
- *Inventory*, which in production applies to raw materials and manufactured goods. Inventory at an insurance company might be too much computer equipment, or insurance products that cost more to maintain than they generate in revenue.

Wireless remedies virtually every type of muda, as we show. But Wireless and Web 2.0 tools have created some fascinating and colorful new species of muda, which desperately require the practice of continuous improvement to keep those tools focused and productive.

## iMuda

The Lean camp describes one of the seven wastes as defects, not only of product but of process. Defects of process are an unfortunate by-product of any new technology. Another is *overprocessing*, or doing more than is required by the customer. Let's take a look at some of the defects and forms of overprocessing born of Wireless and Web 2.0, and call it *iMuda*.

One form of iMuda is the replicate data which is copied and stored in multiple silos, well beyond its useful life (or legal limits, depending on regulations such as Sarbanes–Oxley, Gramm–Leach–Bliley, or HIPAA). Think of a large document such as an engineering document, stored on a server, and in both an e-mail outbox and an e-mail inbox, or a patient record that was destroyed as required by HIPAA, but a copy of which remains on a doctor's hard drive.

Another form of iMuda is the improper application of tools. British Telecom bought 26,000 Panasonic Toughbook computers for use by its linemen, a seemingly natural choice for use on a telephone pole to access engineering and grid drawings. But the workers required two hands to climb and two hands for repairs, and the screens were not usable in daylight.

There is also the iMuda of risk, associated with improperly secured wireless transmissions, and the risks associated with temptation and slacking off. This is more than just a productivity concern: it's also a security risk. In the 1970s, one might disappear into the men's room with a copy of *Sports Illustrated*. Now, one can read *Sports Illustrated* online at the desk, using company equipment, and exposing the company to the security risks of Web surfing. *Sports Illustrated* is a perfectly legitimate enterprise, but the Web sites of such legitimate enterprises as L.L. Bean, *USA Today*, and the Miami Dolphins Stadium have all been infected with malware at one point or another. That fellow in the men's room may not have been productive at the time, but he did not expose his company to the risk that his twenty-first century counterpart does.

It is not enough to simply ask employees to restrict themselves to business use on business equipment; they have never done so. An employee 30 years ago might type a personal letter on an IBM Selectric typewriter, doodle with drafting tools, or sit on the glass of a photocopier and make a dozen copies, but still, the risks were minimal. Also true, Employee 2.0 is used to using computers for entertainment, and so the lines between work and personal use blur, and a remote worker, particularly one who travels, simply cannot be forbidden to use a company-issued mobile computer to manage finances or correspond with family.

## The iMuda of Complexity and Overkill

Continuous improvement favors simplicity as the shortest route to customer fulfillment. In his book, *The Paradox of Choice*, Swarthmore professor of psychology Barry Schwartz opined that individuals may demand choices and features that signal freedom, self-determination, and value, but "clinging tenaciously to all the choices available to us contributes to bad decisions, to anxiety, and dissatisfaction... ."[2]

Technology has put limitless choices in front of the consumer, and in front of the enterprise. "The convergence of media—computer, telephone, TV, film, camera, and the iPod—has put a global marketplace and an abundance of information and entertainment within close reach," wrote business journalist Sol Hurwitz.[3] But although consumers view more choices as being of greater value, about a third of them feel overwhelmed by choices. Still others feel that their attention is divided by multitasking: talking on a cell phone while surfing the Net or text messaging. Furthermore, too many features in a high-tech product can contribute to "feature fatigue." The University of Maryland National Technology Readiness Survey discovered that 56 percent of those who purchase high-tech products feel unable to cope with the overload of features.[4]

Still, technology thunders on, and this is where continuous improvement expertise is vital. Schwartz foresees a growing dependence by consumers upon simplification as a service and filtering aids, for example, people who will wade through your 401(k) options for you, or an online questionnaire that helps you to select a physician from the network of thousands, based on criteria such as years of experience, school of degree, and patient satisfaction.

The business world has long made use of such expertise, for example, by single-sourcing; asking an enterprise software vendor (such as SAP or Oracle) to select ancillary software applications, servers, workstations, and so on; or by hiring a third-party integrator or project manager to handle such an implementation. What this creates is a profusion of experts. In Wireless technology, single-source expertise may come from:

- A Wireless hardware provider
- A Wireless software provider

- A full-service Wireless integrator, in partnership with hardware and software providers
- An enterprise software provider
- A third-party Wireless integrator

"Beware of the cowboys," industry sage Erik Michielsen told us in 2005, when the Walmart mandates (see Chapter 1) created a wellspring of new consultants. "RFID is the next big thing. Everyone who was doing ERP* before and process consulting is all of a sudden an RFID consultant. There's a certain value in getting outside help, but you have to make sure it's qualified and right for the task at hand."

The proof in Wireless is in the résumé; a successful RFID implementer has a number of successful installations from clients who will provide references, and who belongs to trade associations such as AIM Global and CTIA. Still there is no licensure, not even a technical degree that covers the breadth of Wireless offerings, that is, designing a single system that integrates GPS, RFID, remote sensors, and mobile computing. Even the respected RFID+ certification (which *Thin Air* co-author Louis Sirico helped to develop) focuses only on RFID, a subset of Wireless offerings. Arguably, Lean consultants are very well equipped to take into account the breadth of technology, by treating each technology as an individual tool that can be plugged into the value streams.

Professor Shoshanna Zuboff of Harvard Business School and James Maxmin, a former CEO, expect future enterprises to be more dependent upon support than transactions. In their book *The Support Economy*, they describe what they call "federated support networks": whatever the source, a consumer or customer has exactly one place to go for purchase and ownership experience. Dell is a small-scale example. If you buy a Dell computer, you phone Dell for support for the computer, but also for the Microsoft applications on the computer.

The Wireless enterprise requires federated support from selection through ownership; but this must be filtered through the lens of business processes. And that calls for expertise in honing end-to-end business processes, which is the value proposition that Lean consultants offer.

---

* Enterprise Resource Planning.

## The iMuda of Choice

In *The Paradox of Choice*, Schwartz advised that marketers, in designing a customer experience, limit choices to those that matter to the customer. Give them fewer offerings which are easy to choose, and that will foster more satisfactory choices and decisions. This principle translates wonderfully to the production plant, warehouse, or office, and is essential to the Lean goals of satisfying customers and doing the job right the first time.

Consider the paint booth at a Saturn auto plant. The technician does not have a choice of paint colors. Rather, he receives a work order for a specific color to fulfill a customer's order. Any time spent making a choice between one job or another delays both jobs and creates the risk of making a bad choice, such as choosing the easier or shorter of two jobs, versus the one that is behind schedule.

"Choice" is an increasingly loaded word. (Where co-author Dann Maurno was labeled a "brat" by more than one British teacher, modern educators are more likely to tell a child he or she has made a "poor choice.") Choices come in both "good" and "bad" varieties, with high and low stakes. Unfortunately, the stakes are far higher than they were 25 years ago, as we show in the following "horror tales."

## True Tales of Bad Choices: The Super Bowl Virus

Recall our earlier example of reading a copy of *Sports Illustrated* on company time. That's a bad choice, but beyond lost productivity, it represents little threat to the company. Yet employees who made a similar bad choice in 2007, using the Internet, cost U.S. enterprises in excess of $1.1 billion in reimaging infected computers.

What happened? The Dolphins Stadium site had been infected with a keylogging malware, which allowed the attacker full access to any computer, business or personal, that was used to reach the stadium's site, which, just before Super Bowl Sunday, included millions of work computers.[5] In tallying the damage, consider that:[6]

- IT administrators need on average 4.5 hours to reimage a computer, at an average hourly labor rate of $42.50, or $191.25 per computer;[7]
- Employees lose an average 6.5 hours of productivity by being offline, at an average labor rate of $40/hr or $260.[8]

The total cost per incident per employee tallies up to $451.25. Assuming that only a third of the 748,000 average visits per week to the stadium site were from business computers, the incident represented a $1.1 billion business interruption loss, and that's a conservative estimate.[9] Visiting the Dolphins Stadium Web site from work computers was more than simply a "bad choice," it was a costly exposure and a security threat.

## True Tales 2: The Accidental Spy

Sometime in late 2007, New Zealander Chris Ogle ambled into a thrift store, spent $9, and purchased U.S. Department of Defense secrets regarding Afghanistan and Iraq.

This breach in security was as much a surprise to him as it was to the Pentagon. On a trip to the United States, Ogle had purchased a used MP3 player in an Oklahoma thrift shop. Not until he plugged it into his computer, nearly a year later, did he find mission briefings, logistics information on weaponry deployed to Afghanistan and Iraq, and the home addresses and Social Security numbers of thousands of U.S. soldiers. "The more I look at it, the more I see, and the less I think I should be," Ogle told New Zealand media outlet TVNZ.[10] This was not an isolated incident. Two years earlier, the Department of Veterans Affairs lost a laptop with personal information on millions of U.S. soldiers, and hard drives with classified military information have been found for sale in street markets in Afghanistan.[11]

## The iMuda of Risk 2.0

Choices represent risk, which is the unfortunate reality of the diverged consumer/business interface.

Consumer technologies such as social networking and games make their way into the business world, unvetted and unchecked; leisure use threatens business systems. This is new. An '80s-era Sony Walkman cassette player or a copy of *Sports Illustrated* represented no significant threat to an enterprise.

Not that these consumer applications are perfectly useless, quite the opposite in fact. Consider that Web 2.0 social network application Facebook hosts fan pages for Ford Motor Company (with nearly 40,000 fans as of this writing), which provides customer-generated marketing for

Ford. Also true, the Obama campaign made effective use of Facebook for fund-raising, as we show later in Chapter 4. But Facebook also hosts the fan page of a musician named Punch Drunk and an application called "Hockey Fan" for browsing player stats and data. United States companies lose $178 billion annually to casual Internet surfing. If, at a company of 50 employees, each uses the Internet casually for just one hour per week, and the average salary is $20 per hour, then that company loses $10,000 per week and $50,000 per year.

The U.S. Marines take this loss of productivity and security seriously enough to ban the use of social networking sites (SNS) not managed by the DoD. Such sites are a "Proven haven for malicious actors and content," wrote the Corps in a release authorized by Brigadier General G. J. Allen.[12] "The very nature of SNS creates a larger attack and exploitation window, exposes unnecessary information to adversaries and provides an easy conduit for information leakage.... Examples of Internet SNS sites include Facebook, MySpace and Twitter."

Content filters cannot prevent employees from surfing the Web; they can use their personal equipment (such as iPhones) to waste time on Twitter or Facebook, which limits the security risk to the business network. However, if the employee uses that equipment to connect to the corporate network, it is no longer simply a time suck; it is a hole in the corporate firewall.

"Enterprise 2.0" with its 24-hour Internet connectivity, Wireless computers and PDAs, and Web 2.0 marketing, is no longer a four-walls network: it has no walls at all. Enterprise 1.0 was able to secure its four-walled cabin, but is now tasked with protecting hundreds of acres of "prairie," without fences. That's the price of allowing remote workers 24-hour access to enterprise servers, and allowing customers 24-hour access to Web sites and portals.

Some startling statistics that Web and data security provider Websense found in late 2007:

- Eighty-seven percent of e-mail messages are spam.
- Sixty-five percent of unwanted messages contain malicious URLs (including links to spam and malicious sites).
- Fifty-one percent of Web sites with malicious code are legitimate sites that have been compromised, rather than sites specifically commissioned by hackers.[13]

Similarly, the choice to e-mail an Excel spreadsheet including employee social security numbers is, frankly, a lousy choice, but appears perfectly innocent. An employee may have a perfectly legitimate reason for doing so, such as communicating information from a remote operation to headquarters. But it has risks, and the company would be safer if the choice to communicate that information via e-mail were removed altogether. Or, going even further, to eliminate the choice to copy that information to a flash drive or CD (recalling the MP3 player with sensitive DoD logistics files), thus keeping the only copy of the information on the controlled network. This is by no means tyranny, this is loss prevention. And employees understand that.

## The iMuda of Terabytes

Data by itself is not a waste, but there are ways to gather and use data wastefully. "Data gathering has to get done," said Toby Rush, president of Rush Tracking Systems. "If you don't do it automatically, then someone must do it." And usually, manual data collection is a waste of someone's time.

Furthermore, data can be presented incomprehensibly, and therefore, wastefully. Business information (BI) applications are supposed to make that information digestible and easy to use. But author Ralph Kimball wrote in his book *The Data Warehouse Toolkit* (second edition), that BI reporting tools, "as powerful as they are, can be understood and used effectively only by a small percentage of the potential data warehouse business user population," which does not include the executives who need the information.[14] Analysts describe an information "pyramid," in which 10 percent of users qualify as advanced users, chiefly analysts and IT people, and the remaining 90 percent are nontechnical business users such as C-level, VP-level, and operations-level executives. As a whole this group finds BI tools too complex to use, and opts for familiar sources of intelligence instead (chiefly, Microsoft Excel, the hands-down favorite BI tool among that 90 percent). Excel has its charms: it's easy to use and configurable. But it is limited in that it is not designed as a tool for real-time data mining and application integration.

Finally, data may be collected without purpose. As SAP's Claus Heinrich described in *RFID and Beyond*, "Millions, if not billions of points of data could be collected, but not all information is created equal. Executives ... must first focus [upon] information that will make the most difference," and then seek the incremental improvements in process.[15]

Before collecting data, advised Heinrich, an enterprise must ask:

- What important information is unavailable?
- What information could suppliers and partners find valuable?
- How will processes have to be changed?
- What new metrics can be monitored?

To improve processes, the enterprise must then ask, among other questions:

- Where are the errors and delays?
- What automation does the better information make possible?
- What processes have bottlenecks?
- What partner and supplier processes can be improved?[16]

At this point, the enterprise can apply a data collection method, and a process improvement.

Another option is to seek new methods of improvement. Although Wireless appears to be, simply, a tool, let's not forget that the telegraph was also simply a tool. But it was the first tool to allow one-to-one communication by people who could neither see nor hear each other first hand. It, and the telephone, fostered one-to-one communication that was impossible before, and as a result created more communication.

Similarly, "More information at first improves the current way of doing things, and then creates completely new ones," as Heinrich wrote.[17] For example, RFID at its simplest identifies what is present: perhaps a big-screen TV that arrives on an outbound shipping dock. The item changes from pallet to pallet, but the location (the loading dock) is fixed. RFID evolved to track the location of an item, even its movement, using real-time location systems. Finally, RFID enables us to conduct inventories, recall products, charge drivers at toll plazas and gas stations, and so on.

Still, the data a company collects must be useful. Heinrich predicted that the real-world awareness of enterprise and supply chain data will require "an expansion of data warehouse, data mining, Online Analytical Processing (OLAP) techniques" and other methods of managing terabytes of data.

## OVERCOME THE INERTIA OF LEGACY

One challenge to the advance of Wireless is overcoming the inertia of legacy. In other words, companies tend to stick to applications that have worked well, such as bar codes, wired telephony, and wired LANs.

Why would a company trade its wired telephones for a wireless system? Wired phones have worked perfectly well for nearly a century, and typically provide sound quality on par with VoIP and better than cellular signals. Similarly, a company would find it difficult to justify taking down a LAN that still does its job.

Similarly, replacing a bar-code system that works perfectly well in a supermarket with RFID seems absurd. Yes, every item could be tagged with RFID and priced automatically, and without a checkout clerk. But the reasons for keeping the clerk include that (1) clerks, passing every item through their hands, ensure that every item is paid for; (2) those automatic checkouts with bar-code scanners are far from 100 percent reliable, and clerks rectify the errors; and (3) supermarkets and retailers employ hundreds of thousands of Americans, a good many of them young people in their first jobs. Most RFID integrators (ourselves included) will recommend bar codes when bar codes make more sense.

Still, a well-working tool is not necessarily the best tool. The tried-and-true nature of bar codes can stifle advancement in manufacturing, suggest Wireless experts Michel Baudin and Arun Rao in their treatise "RFID Applications in Manufacturing." Although bar codes have improved data input productivity and quality over manual keyboarding, "The best opportunities for RFID are where bar coding is so pervasive that production operators spend a *perceptible fraction** of their time scanning bar codes ... [such as in] computer assembly, for pick validation and component serialization."[18]

Bar codes require manual manipulation, an unobstructed line of sight, a clean and undamaged label, and a speed low enough for one-at-a-time scanning. A passive RFID tag the same size as a bar code requires no line of sight, can be read from distances up to 15 feet away, at a rate of hundreds per second (in some cases), and can hold significantly more data, up to 64 K. Compare that to the single Unit Product Code, or UPC, of a bar

---

* Italics ours.

code. Finally, an RFID tag eliminates that element of human error that keeps a process at the Three Sigma level.

Still, RFID is costlier than bar codes, which cost essentially less than 1¢ to print, practically free. Passive RFID tags cost a dollar apiece in 2004. In 2009, a better-performing tag with more features cost 10¢ in quantities of 10,000. But 10¢ is still too expensive if the return is not equal to or greater than the business pain that it relieves.

Yet because of their inherent limitations, there are limits to the pain that bar codes can relieve. Recall as we saw in Chapter 1 that the International Air Transport Association (IATA) estimates the accuracy of bar-code read rates in baggage handling of 80 to 90 percent, versus 95 to 99 percent for RFID. At those figures, the airline industry could reduce lost or delayed bags between 12 and 15 percent and save $760 million per year with RFID baggage tracking.[19] That is consistent with results from, for example, Hong Kong International Airport, McCarran International Airport in Las Vegas, and London's Heathrow* Airport, all of which now use RFID baggage tracking. The return on investment more than surpasses the expense to implement RFID and the ongoing operational costs; RFID passes the tests of cost/benefit analysis, handily.

Appendix D, "Lean Wireless ROI [return on investment]," describes methods of cost/benefit analysis for Wireless technology. No responsible CEO or operations manager will allow a Lean Wireless transformation, without being certain that (1) it pays for itself and (2) it does not disrupt business as usual, rather, it allows more business than usual. A good many of the benefits, for example, customer satisfaction and retention, are not easily quantifiable, but a good many returns are also measurable and immediate. Recall that fully 75 percent of companies using RFID expect a return on investment within 18 months, according to an ABI research study.[20] They have every reason to expect that. As we show in Chapter 4, the Mercy Medical Catheterization Laboratory saw a 500 percent ROI in an RFID system in 18 months. That return included reducing inventory levels by $376,587 (25 percent) in the first half of 2008, and reducing waste from expired products by 40 percent. Mercy Medical also saved 1.5 hours per weekday, or 7.5 hours per week, in conducting manual inventory.

Mercy Medical began its implementation in late 2006. As recently as 2004, the HDMA Healthcare Foundation published a pessimistic cost/

---

* As of January 2010, the largest RFID installation in the world.

benefit analysis of using RFID which showed investments in the tens of millions of dollars, and payback periods in years, not months (which at the time, may have been accurate).[21] But in two years, the technology had standardized, tag costs went down, and Mercy Medical's director of Six Sigma revisited the technology. Both McCarran Airport and Mercy Medical had working systems in place that did the job, but chose to overcome that inertia of legacy.

However many success stories there are, Wireless providers "Need Lean," says Dr. Can Saygin of the University of Texas, San Antonio's Center for Advanced Manufacturing and Lean Systems. "If technology developers were to approach industry from a lean productivity standing, there would be more [acceptance] of the technology. You have to first do some cleaning in house, understanding the process, value-stream mapping, and then find areas to embed [technology]."

## FOCUS FIRST UPON INTERNAL VALUE STREAMS

Perhaps that is where Lean and Wireless diverge most sharply as solutions. Wireless is not as customer-centric as Lean's value stream model; rather, it is work and task-centric, which ultimately satisfies everyone:

- The employer
- The employee
- The customer

To the employer, the value of Lean Wireless is in competitive advantage, ease of ownership, higher profit, and a lower cost of goods sold, which ultimately also benefits the customer. To the employee, the values include workplace satisfaction, career fulfillment, and ongoing employment. No one has ever whistled while he performed scut work, which Wireless and automation eliminate. To the customer, the value is in receiving what he wants, when, where, and how he wants it, as Womack and Jones described in *Lean Solutions*.[22]

The Lean theory of productivity is to the point and pragmatic, which is why it is so powerful. It asks, "What creates (or doesn't create) value for a customer?" and aligns the value-add activities along a value stream. The value-stream, customer-focused approach ensures that any function along the value stream (e.g., production, logistics, service) has a higher priority than any process outside the value stream (e.g., HR, the company cafeteria, or the security desk). After all, the customer is invested in the qualifications of a field engineer, but not in her dental plan or lunch habits. The customer, in essence, "outranks" everyone in an organization. Processes along the value stream, such as sales and production, take precedence over functions outside it.

Lean tends to overlook the enterprise as a customer, when the enterprise is its own and biggest customer. The enterprise value stream includes such functions as HR, security, accounting, maintenance, and governance, risk, and compliance (GRC), among dozens of other functions. "Companies put a lot of effort into optimizing the value stream around product," said Motorola's Prouty, "but I would say that 80 percent of the potential savings is off the value stream. Accounting, maintenance, they all add to the cost of goods sold." Interestingly, 80 percent is exactly the figure that most Lean sources quote as being "off the Prado chart," or not adding value to the customer.

"I once had a quoting department report to me," said Prouty, "and we guessed what our labor was," for billing. The result was either underpricing or overpricing the production labor costs. "Either way you lose money. If you're overpriced you lose bids and price yourself out of the market; underprice it and you lose money. Value stream mapping never sees that.

"Toyota looks at the value stream map across the entire company, for example, taking into account regulatory issues, collecting material safety data sheets and checking fire extinguishers." Toyota automates a good many of those functions, thus, an MSDS audit might take an hour versus a whole day, but the improvement would never show up in a traditional value stream. So the need is to broaden the value stream. You need to recognize internal value streams, value them highly, and manage them with the same priority as a customer value stream, rather than dismiss internal processes as "waste."

No place is farther off the customer value stream than a company cafeteria, but Motorola implemented bar codes in its company cafeterias long before it was common. A diner would swipe a card and a departmental account was debited. The company collected immense data on eating habits and what was popular. "If you look at that from the perspective of the value stream, it doesn't show up anywhere," said Prouty. "But it helps us make employees' eating experience more efficient; they spend less time waiting in line to pay for lunch, and more time actually taking a break, and not coming back late."

A few minutes here and there, of course, add up. Let's look at a simple but real-world example. Raytheon has long been one of the world's most significant defense contractors. In the mid 1980s, the Andover, Massachusetts plant employed 7,000 people, giving each 45 minutes for lunch in a vast company cafeteria. Had 7,000 people been obliged to eat offsite, each would have required a full hour (at least) to navigate the colossal parking lot, stand in line at one of the town's few restaurants, and return to work. That savings of 15 minutes per person multiplied by 7,000 people per day enabled Raytheon to recover

- 105,000 minutes/day of productivity or
- 219 work shifts/day, the equivalent of hiring 219 full-time workers

Certainly it cost Raytheon to run a company cafeteria, but even if the cafeteria ran at zero percent profit, the net recovery is 219 work days of productivity, every day; and about 210 years of productivity, per year, recovered by cooks, servers, and cashiers. Those savings drove down the cost of the Hawk and Patriot missile systems that were created in the value stream and sold to customers such as the U.S. Army who, presumably, did not care where Raytheon employees ate their BLTs but cared about how much a missile cost.

Let's say that job functions such as HR, security, and a cafeteria are "on the banks" of the customer value stream. These are the functions that are outsourced, eliminated, and picked over for cost savings, yet through automation, they represent an opportunity for enormous cost recovery.

## IDENTIFY OPPORTUNITIES FOR WIRELESS VALUE

They may not speak the same lingo, but an adage that both Lean and Wireless experts use is "Information replaces inventory." Both practices aspire to minimize inventory, or eliminate it where possible.

Information does indeed replace inventory, as we saw in the case of American Apparel. But in Wireless best practices, information also replaces:

- Routine tasks, such as counting inventory
- Routine paperwork, through automation
- Routine decisions (including bad decisions)

- Cash transactions (think of the E-ZPass toll collection, which uses RFID transponders)
- Physical mailers (in the form of e-mail, instant message, voicemail)
- Security checks and identity verification
- Keys
- *Genchi Gembutsu* (go-and-see) of equipment checks
- Expenses of air and land travel in gathering for kaizen or quality circles
- Physical wired infrastructures (telephony and networks)
- Log-ons
- Phone calls
- Meetings

Every such replacement represents a cost savings and a triumph of continuous improvement.

Replacing all of those processes with information calls for a higher level of information than simply "enterprise data." It calls for what SAP CEO Claus Heinrich described as Real World Awareness, or RWA. As he defines it, RWA is "The ability to sense information in real-time from people, IT sources, and physical objects—by using technologies like RFID and sensors—and then to respond quickly and effectively."[23] RWA Wireless originates on the edge, wherever the work is done, be it at a machine, on a forklift, at a patient's bedside, or in a truck. RWA is the movement from manual, batch-oriented, and limited information to always on, always aware, always active networks, running in the background, enabling us to do fulfilling work.[24]

## Value from Mobile Workers and Equipment

How often do you reach a co-worker on the first ring? Even supposedly "deskbound" workers are more mobile than they think. A worker can be considered "mobile" if he spends 25 percent of the time away from his workstation (or has no workstation at all). Rather than thinking of workers as either deskbound or "road warriors," let's look at a newer paradigm from, of all places, *Mobile Workforce for Dummies*, which describes four types of workers based on their mobility:[25]

- *Deskbound workers*, who tend to work in back-office functions such as accounting or administration.

- *Teleworkers*, who work from a fixed location offsite (frequently from home).
- *Campus nomads*, who wander from meeting to meeting, group to group, building to building. They may use cordless phones, cell phones, or Wi-Fi networks in order to be mobile yet still connected. (Dell refers to these individuals as *corridor warriors*.)
- *Road warriors*, including salespeople, delivery drivers, and field service personnel. Executives who travel frequently fall into this category.

Campus nomads and corridor warriors may be the norm in a modern enterprise, and they are busy with far more than just meetings. This category also includes:

- IT technicians, who are out and about performing upgrades
- Maintenance, janitorial, building, and grounds personnel
- Production operators
- Security personnel
- Doctors and nurses

Most companies begin by equipping salespeople and other "road warriors" with mobile and Wireless technology—typically a laptop or BlackBerry—but ARC's Ralph Rio advises that a better approach is to simply observe who moves around both on campus and off and shouldn't be tethered to a computer. With Wireless input/output (I/O), a machine operator need not visit a machine to record its data, and the quality of the data is more likely to be correct. And a machine operator who is equipped with a Wireless Communication device can be responsible for more than one machine, if she manages by exception and is alerted to those exceptions on a PDA.

One of the key Lean benefits of a Wireless sensor network is in process optimization, but also in waste avoidance: of labor, defective product, and of breakdown versus predictive maintenance.

Let us consider the continuum of manual sensing, to wired sensing, to wireless sensor networks (WSNs). The equipment on the network can be a boiler, steam turbine, bioreactor, patient, or other critical application.

1. *Manual sensing and manual communications.* This is "making the rounds," reading the dials, taking the temperature, and jotting it

all down. This is the most basic and labor-intensive configuration. In the case of a boiler, an operator travels to a piece of equipment, inspects it, manually records the results and measurements (such as pressure), and travels back to some control center. Ideally, that person is very knowledgeable regarding the equipment. The more frequent the inspection, the more costly it is.

2. *Wired sensing and manual communication.* This is an improvement over #1, but equally labor intensive. One or more sensors are built in or added to the equipment. An operator travels to a piece of equipment, checks the sensor(s), manually records the results, and travels back to the control center.

3. *Wired sensing and wired communication.* Here's where the labor savings begin. One or more sensors are still used, but the operator no longer travels to the equipment to take sensor readings. Monitoring is more frequent and continuous. This type of sensing is common for operation-critical equipment.

4. *Wired sensing and Wireless Communication.* One or more sensors are still used, but the wire is replaced with a radio frequency (RF) transmitter. The transmitter uses a constant power source, typically an electrical outlet, but solar-powered sensors are becoming more common in areas where wiring is difficult.

5. *Wireless sensing and Wireless Communication*, or, completely airsourced. This is the most sophisticated configuration. One or more sensors are still used, and only the RF transmitter is part of a sensor device.

As we noted above, #1 is the most costly, not only because of the labor involved, but also because of the risk and maintenance. Infrequent inspections and slow problem response time are unacceptable for important or operation-critical equipment. In properties with thousands of pieces of equipment, it is common to inspect each asset once a month or even once a quarter. If something fails, the technician may not know until the next inspection and would have limited information to diagnose the problem for prevention in the future.

The completely airsourced configuration of Wireless sensing and Wireless Communication is by far the most cost effective. Recall from Chapter 1 the U.S. Department of Energy's estimated installation costs of $200 to $400 per sensor plus wiring at $200 per foot. This configuration provides continuous monitoring without wires, and without the labor of making the rounds.

Moving into materials and equipment, the more mobile a machine is, the more likely that Wireless will add value. Two very real examples are (1) in hospitals, which are beginning to use real-time location systems to track crash carts; and (2) in equipment rental lots, which use RFID to track the presence and location of pieces of equipment. This eliminates the need for redundant equipment, and maximizes the turnaround (and profit) of an individual piece.

## Create Job-Specific and Micro-Value Streams

Continuous improvement experts are employed to identify macro-value streams in an organization, usually following the direct route from order to cash. That's logical. One example would be Boeing delivering a 787 Dreamliner jet to U.S. Airways. Once they have established the macro-value stream, continuous improvement experts will focus upon the micro or drilldown value streams; for example, the most cost-effective and fastest way to ramp up a particular piece of key equipment.

Wireless requires a far more granular approach. The forward opportunity in Lean Wireless is to define job-specific value streams, identifying an individual's core and noncore processes, value-add and nonvalue-add (NVA) activity, then automating and airsourcing all noncore decisions, judgments, and tasks. This frees a worker to create value.

Let's look at a pharmacist as an example. Presumably, the order-to-cash value stream involves receiving medications in stock, receiving a prescription from a physician, then filling that prescription for a customer in need. How much of a pharmacist's job description actually flows along that value stream? In the course of a day's work, a pharmacist must:[26]

1. Ensure that prescription drugs are dispensed properly to ensure patient safety.
2. Counsel customers on the proper use and storage of their prescriptions.
3. Communicate with doctors and other healthcare providers to renew prescriptions and discuss drug protocols and potential drug interactions.
4. See that insurance forms are properly completed, signed, and filed.

5. Manage inventories efficiently to avoid out-of-stocks.
6. Constantly monitor sell-by and expiration dates to ensure that out-of-date drugs are not dispensed.
7. Manage timely return of near-expiring drugs to ensure maximum manufacturer rebates.
8. Take full responsibility for the security of Class II narcotics and other controlled substances to prevent theft or fraudulent dispensing.
9. Properly supervise the pharmacy technicians assigned to restocking and other routine duties.

Only tasks 1 through 3 involve medicating patients, and put a pharmacist's unique skills to work, and only tasks 1 through 3 treat the patient as a customer. Tasks 4 through 9 are certainly consequential, but treat the enterprise (the retail pharmacy or hospital) as the customer. But these tasks are simply a poor use of a pharmacist's specialized skills and knowledge. It is simply a distraction to make her responsible for double-checking expiration dates, or requiring an assistant to do so, when that task can be airsourced.

Similarly, in automotive production, painting a car is a value-add task, but filling out a work traveler (the paperwork that travels with a job) is a waste of a paint technician's time. In healthcare, taking a temperature is a value-add task, but recording the result is a routine task that can be automated.

In essence, if the macro-value stream is like a river, the job-specific value streams are like streams and rivulets.

## Apply Airsourcing as You Would Outsourcing or Offshoring

Airsourcing is much like outsourcing and offshoring in that it eliminates repetitive, nonvalue-add, or lower-skill work.

Consider outsourcing first. An apparel manufacturer specializes in styles, not in benefits administration, and so may outsource that important but noncore function. Similarly, the manufacturer may see its core function as clever clothing design rather than manufacturing, and offshore the manufacturing to Taiwan or to Mexico. This is the way that most apparel manufacturers work now; you will not find a Nike or Chico's plant in the contiguous United States. For that matter, the company can outsource its designs, so that a clothing "manufacturer" operates out of a suite of offices in Manhattan and doesn't directly employ a single production worker.

The last decade has seen a tremendous advance in outsourcing and offshoring, from production and into services. Everyone is familiar with the Bangalore- and Manila-based help desks of, for example, Microsoft and Netgear, but a good deal of the news you read originates there as well. In *The World is Flat*,[27] Thomas L. Friedman described the outsourcing of journalism to India. Online research of a company's sales data can be done as competently, and far less expensively, in Bangalore as it can on Wall Street, and "top-ten journalism" ("Ten Embarrassing First Date Gaffes!") requires research skills more than journalistic skills.

The point of airsourcing is to eliminate nonvalue-add tasks from individual jobs, rather than to eliminate the jobs themselves.

## Return to Processes, and Mobilize Them

Mobilizing a process creates an opportunity to troubleshoot the process; Six Sigma, after all, calls for objective data and standardization in perfecting a process, and Wireless enables both.

"The reason a lot of bad processes happen in the first place is that people develop coping mechanisms around ambiguities, and those coping mechanisms become institutional wisdom," observes Toby Rush, CEO of RFID solutions provider Rush Tracking Systems. "Safety stock is because you don't know your inventory. Pad time is because you don't know how long a process takes." Both safety stock and buffer time are, of course, considered waste in Lean. "RFID creates confidence in the system and accuracy, with realistic, objective metrics," said Rush. "Management is able to measure performance and get visibility when movement happens."

Wireless further enables a company to manage mobile assets with refined business processes embedded into a company's IT system. In particular, those processes that we looked at earlier that are nonvalue-added but still required (such as accounting, security, compliance) can usually be perfected, then automated and streamlined with mobile technologies.[28] The benefits to this approach, as ARC Advisory Group defines them, are that:

- Mobile resources are better managed.
- Other business processes become more inclusive and responsive.

- Lean manufacturing has new opportunities to remove waste.
- Six Sigma's design/measure/analyze/improve/control (DMAIC) process gains access to more data for analysis.
- Edge visibility (connectivity to where work is done or customer value is created) provides new opportunities to eliminate waste and head off defects (both Lean objectives).

## PLUG VALUE STREAMS INTO THE "INTERNET OF THINGS"

The much-ballyhooed Internet of Things would essentially treat everything, and everyone, like a URL address; on the IOT, you can determine something's location, determine its condition, and control it over the Internet. (We say "can," because the earliest iterations of the IOT already exist.) The IOT does not apply to "smart assets" alone, such as computers and PDAs, which have computing power or wireless capability built in; it can also apply to an unwired "dumb asset" such as a clothing item, stapler, or bin of parts, if some intelligence (such as an RFID tag) is added to the asset to connect it to the IOT. Recall our example in Chapter 1 of clothing items in an American Apparel store. A pair of leggings becomes traceable onscreen by affixing an RFID tag. A bioreactor connects to the IOT through a remote temperature sensor. A truck connects to the IOT through its GPS device. And a doctor (who for all his education is an unwired "dumb asset") connects to the IOT through an RFID-enabled badge. The Lean implications are enormous; both smart assets and dumb ones can be tracked, controlled, automated, and put to work in business processes. That's why industry has driven the creation of the IOT, with sponsorship and standards participation.

The term "Internet of Things" was coined at the Auto-ID Center at (where better?) the Massachusetts Institute of Technology (MIT). Englishman Kevin Ashton was then a brand manager at Procter & Gamble, and co-founded the Center in 1999 along with MIT professors Sanjay Sarma and Sunny Siu. Ashton was reacting, with disgust, to finding shelves empty of P&G product at retail outlets. Bar coding did not provide sufficient analytics. Greater intelligence was required at the edge, which RFID could provide.

The idea caught fire and has burned brightly ever since. Procter & Gamble, Gillette, Kraft, Unilever, and Walmart among others (mostly retailers and consumer goods manufacturers) ponied up $20 million to fund the Center to solve the problem.[29] The Auto-ID Center evolved into the Auto-ID Labs Network, which encompasses more than 100 corporate sponsors and seven research universities worldwide; Ashton moved on to co-found ThingMagic, an RFID company located just outside MIT; and the Auto-ID Center, along with EPCglobal, worked together to create global standards for RFID and then licensed those standards to the international GS1 standards body.* So, a decade later, do we have the Internet of Things?

Yes, but it has moved beyond the Auto-ID Center's vision, and spread well beyond consumer goods, manufacturing, and MIT. And, it has spread well beyond RFID.

## The Intent of the Internet of Things

Kevin Ashton spoke to us ten years, almost to the day, after he coined the phrase "The Internet of Things," a phrase that he said somewhat ruefully "is my lasting legacy. I'm definitely the world's authority on what I *meant* by that, and it gets misunderstood. Like anything that gained currency, it means what people want it to mean." Over the past ten years, he has observed the term "Internet of Things" become a fuzzy-edged catch-all term for ubiquitous wireless connectivity.

Ashton's vision of the IOT is a world wherein the Internet (which is a network of bits) captures and stores and transmits information about the physical world, without human help. Given his roles at Procter & Gamble and at ThingMagic, he conceived the IOT as a supply chain application. The Center would in turn hammer out the RFID standards adopted by Walmart and Sam's Club.

As Ashton described his vision of the IOT, "If a shipment of oranges arrives at a distribution center, the ideal is that the DC knows those oranges have arrived. They know how many there are, know their

---

* GS1 is, to quote its literature, "A global not-for-profit organisation dedicated to the design and implementation of global standards and solutions to improve the efficiency and visibility of supply and demand chains globally and across sectors." The GS1 system of standards is the most widely used supply chain standards system in the world.

sell-by date, and know their condition, without a human being doing any counting or data entry." In the Internet of Things, machines could capture data, without human support, about things moving around in the world.

That's very valuable—if you can make it real.

If the Internet is a network of bits, then the supply chain is also a network, a series of connections between nodes, which include farms, mines, factories, distribution centers, retailers, consumers, and so on.

## Redirect the Lean Mission from Making Things to Moving Things

As Ashton describes it, "While the original intention of the Lean movement was to make it cheaper to make things, and to be less wasteful so we can be more profitable, a more compelling problem now is to be more efficient in *moving* things." After all, moving things is what distinguishes companies like Amazon, Dell, and L.L. Bean; each differentiated itself as much upon fulfillment as it did with product.

Although the missions of production and fulfillment seem different, both involve delivering quality goods to the end customer, and the supply chain design makes all the difference. Toyota represented lower-cost, higher-quality goods, as did supply chain champion Procter & Gamble. Think what you will of Walmart, but "It is remarkable how they make luxury things like big screen TVs available to people without a luxury income," said Ashton. "That's the miracle of supply chain engineering," and one traceable in part to Wireless technology, specifically, RFID.

An efficient supply chain becomes increasingly vital, because as economists describe it, money "sloshes around the world" and fulfillment must follow money wherever it goes. The average wealth of the world population is growing, which requires a well-oiled supply chain. Jack Perkowski, the auto parts manufacturer profiled in Friedman's *The World is Flat*, remarked to co-author Dann Maurno in a 2006 interview that in his years of doing business in China, he observed a direct correlation between the prosperity of the average Chinese family and the number of televisions glowing from their windows as Perkowski drove by after twilight. This is as good a measure as any, and better than most, of economic conditions in China.

Is our current production-oriented infrastructure and economy up to the task? No, believes Ashton. "The only way that global capitalism in the 21st century works is if the world population is below 3 billion, and only 10 percent of the population can afford shiny new things and the remaining 90 percent is engaged in making them." In essence, if it reverts to the twentieth-century-sized market for which it was designed, and which was nearly entirely dependent upon U.S. consumption; the United States was the hub of the global supply chain, and the supply chain ended here.

Finally, the supply chain model applies to far more than consumer or business-to-business (B2B) fulfillment. As we show in Chapter 4, voting is a supply chain process; until we give over to electronic voting, the United States is entirely dependent upon the physical movement of voting equipment and paper ballots. The Department of Defense treats combat readiness as a supply chain challenge; as we saw in Chapter 1, the United States Air Force's eLog21 initiative is a supply chain transformation, aimed at delivering warfighters to engagements rapidly and effectively.

## The Internet of Computers versus the IOT

A fundamental difference between the Internet of Things and of computers is that a Web page (such as CNN.com or nationalenquirer.com) can be viewed by anybody, anywhere, at any time; an object can be in only one place at any time. Thus, as CEO Ravi Pappu of ThingMagic describes, "There is some inherent locality in the Internet of Things, that is not on the Internet of computers."

The Internet of computers has no connectivity to place; a Uniform Resource Locator (URL, or Web address) is a sort of place where information can be found: information about politics at whitehouse.gov, about economics at Bloomberg.com, and about the Three Stooges at threestooges.com. It does not matter at all to the end user where that information is stored; she retrieves it at the interface.

Another difference is the degree of dynamism. The Internet can tell us historic information about something, for example, about a meeting between President Obama and Venezuelan President Chavez, which has already taken place. Dynamic interaction (shopping, playing games, sending messages, and the like) is driven by humans.

Conversely, dynamic interaction on the IOT is driven by the things themselves, using technologies such as RFID, GPS, RTLS, and remote sensors.

## RTLS and WSNs

The Wireless nerve path, like the one in Chapter 1, is rare, mostly theoretical, but proven already in two applications: real-time location systems and wireless sensor networks.

RTLS has been commercially available since early 2000, enabling enterprises to locate mobile assets and people within a four-wall enterprise. According to the experts at ABI Research, Wi-Fi-based real-time location systems will become an $800 million market by 2012, with healthcare being the strongest vertical.[30]

The Lean benefits of RTLS include:

- Eliminating the waste of searching for assets
- Process optimization, by locating key mobilized equipment and personnel
- Real-time WIP visibility, to eliminate production bottlenecks
- Asset optimization and eliminating the need for redundancy
- Improving staff safety by tracking people in hazardous work environments

The RTLS end-user interface is typically a map view of an area, such as an emergency room or office. Punch in "crash cart" at a hospital and see a dot on a map. Users typically receive alerts however they want them, via e-mail, text message, voicemail, message board, pop-up, or other event-driven communications device. Finally, any Wi-Fi enabled device (PDA, laptop, bar-code scanner), or anything with an add-on Wi-Fi tag, can be tracked with Wi-Fi-based RTLS.

If RTLS is location-centric, wireless sensor networks provide information about condition, using sensors: temperature, acidity, vibration, motion, or oxygen level (or even blood pressure). Analyst firm ON World believes the WSN market is on track to reach $8 billion globally in 2010, with $5 billion from industrial applications. WSN is used widely in process manufacturing such as food or pharmaceuticals, for example, measuring the acidity in a chemical vat.

A few Lean benefits of wireless sensors include:

- Preventing the waste of unplanned failures
- Reducing the cost of manual inspections
- Reducing the structural cost of regulatory compliance or GRC
- Optimized operations through continuous monitoring (and improvement)

## The 5Ws versus 5Ss

Both RTLS and WSNs have been called the Internet of Things here and there, but ThingMagic's Pappu disagrees. "Location is not the only information; identity, location, temperature, pressure, and higher-level constructs—like how many times did this object pass this point?—all of that data is a property of the physical object."

With technologies such as RTLS and WSNs, an asset is capable of telling us far more than what something is or where it is, but also, who has it/owns it/touched it/sold it/stole it, when it was last seen, why it was moved, what it is doing (moving on the hospital floor, being powered up, producing at capacity), and so on. An asset can even tell us with sensors how it is "feeling."

"The information we gather is the 'five Ws' of who, what, when, where and why," said Ekahau's Vice President of Business Development Tuomo Rutanen. "Compare those five Ws to the 5S of continuous improvement," which are Sift, Sweep, Sort, Sanitize, and Sustain; the theory of 5Ss is to clear the field of unnecessary inventory, movement, personnel, and the like, such that what is left over is value-add, and problems are quickly identifiable.

Still, the first and most important parameter is what something is; everything else is either secondary or unnecessary. "Location is never enough," said Ashton. "If I tell you there is an object three feet away from you but don't tell you what it is, the information is not useful."

The challenge in plugging an asset into the IOT is to choose the Ws carefully; all else is a waste of effort and information. If, for example, you have a carton of milk in the refrigerator, it has physical properties such as length, width, and height, and also properties of condition, such as its temperature, and its historic temperature. Did the temperature ever go above or below certain thresholds? As Pappu describes it, "That carton of milk is a member of the Internet of Things, which had nothing to do with location, it had to do with *temperature*, and if it crossed a threshold or not." ThingMagic has deployed systems that measure the temperature of objects periodically when they come within range of a reader, then downloads that data onto the Internet.

"An alternative analogy is a subcutaneous glucose sensor, measuring the glucose of a patient with diabetes. It sends all that data back to a nurse's station at a hospital. If the glucose is dropping, the base station can change the delivery of insulin." In this instance, the glucose sensor becomes another "site" on the IOT, providing condition versus location.

## Why the IOT Is Not the Internet

We said above that the 5Ws must be chosen carefully. This is because 5W visibility into things (knowing all the who/what/when/where/why information) is unnecessary, and wasteful. It is perhaps useful to know who handled a shipment of pharmaceuticals (the chain of custody), where a shipment is, its temperature history, and so on; this knowledge will be mandated in the California ePedigree regulations, which we discuss later in the chapter. The same requirements do not apply to a shipment of car tires. Thus 5W visibility for the sheer purpose of 5W visibility is what young digital natives call a "time suck," an application that takes up enormous time for no purpose.

Any particular thing is of limited interest. If it is a company's thing, for example, calibration equipment or a notebook computer, it need not (and should not) be searchable by the world's population, and at their leisure. What companies require is more of an intranet of things, a secure and closed-loop system. Pappu describes this as a "Reality Search Engine," which searches the space around us in real time. He details three degrees of space:

- *Manipulatory space*, where objects are close at hand (e.g., a supermarket aisle)
- *Ambulatory space*, where objects might be in walking distance (e.g., on a hospital floor or an office building)
- *Vista space*, within eyeshot, such as a parking lot or sports stadium

ThingMagic customer Mediacart produces an RFID-enabled shopping cart that searches manipulatory space for bargains, low-calorie items, and the like. The Ford/DeWALT/ThingMagic Tool Link application queries the ambulatory space of a vehicle. Finally, companies that use RFID to locate vehicles on a rental lot, or theme parks that use RFID bracelets to locate children, query the vista space.

This is more of a challenge than it seems. The IOT is compelling to the Internet generation. "We're so used to the idea that things are for our pleasure; we're the suppliers and customers of computerized information and everything's about us," Kevin Ashton told the authors. The temptation exists to create as much information as it is possible to create, and far more than is useful. The IOT must be more like a virtual private network (VPN) than like the Internet itself.

Whatever the possibilities of both RTLS and WSN, they must be reined in, to some degree, and the IOT must be focused upon useful information. "Again," said Ashton, "what something is is the primary piece of information you need, and information like condition, location, price, origin, age, is it toxic, how hard you can squeeze it and if you do will it explode, is all secondary and tertiary data." Location can be useful, but location within a six-inch square is a far more expensive proposition than location to a particular room.

## Devise Lean, Universal Number Schemes

Two things will be common to all objects on the IOT:

- A unique identifier to distinguish one object from another, akin to a URL address
- The ability to report data or identity over a short-haul wireless link, such as RFID

Two examples that fulfill both requirements are the EPCglobal and DoD numbering systems for suppliers.

The technology is surprisingly the easier of the two requirements. We can attach an RFID tag or remote sensor to practically any dumb object and put it on the global IOT (or an enterprise's private Intranet of Things). A "unique identifier" is another matter.

The challenge is not in the quantity of numbers. A 96-bit EPC number is theoretically scalable enough to tag every grain of rice in the world. "With a 128-bit EPC, you can uniquely identify everything you could ever conceive of being made," said Ashton, "and that work has been done, though not widely adopted."

> Numbering turns out to be a cult-like religion for people who have a number they use and don't want to use another, and there are arcane debates about mine versus yours. In supply chain numbering, you can use a vehicle identification number or a national drug ID number or a GTIN or UPC, and there's an absolute holy war about it.

Cataloging all the numbering schemes in effect today would require years. Each was conceived for a specific purpose and is context dependent; one reason that International Standard Book Numbers (ISBNs) are 10 to 13 digits long is to accommodate bar codes; if ISBNs were 20 numbers long, the bar-code lines would have to be thinner, which would make it difficult for scanners to read them. Requiring libraries and book retailers to purchase high-resolution scanners that can read microscopic bar codes is simply impractical.

In addition, a single numbering convention would be difficult to implement, when different kinds of physical assets are subject to different business rules. Compare the security requirements of a book in a bookstore to a box of M240 rifles on its way to a U.S. Army base. Some corporate assets will require firewall-type security, thus, a private numbering convention. If consumers wish to use the IOT to tag their books, gym bags, and car keys and search for them on the Internet, they will require an open standard, or the ability to devise their own numbering schemes, the same way they name file folders.

Still, we need numbering conventions to stretch supply chain visibility beyond four walls; using a common numbering system (EPC numbers), Walmart and its suppliers can trace a shipment of bicycle helmets along the entire supply chain. (They do not in fact track the helmets; they track the numbers affixed to those helmets with RFID tags, and the movement of those numbers. From the movement of the numbers, they extrapolate the movement of the product.)

Vice President and Chief Marketing Officer Andre Pino of Acsis told the authors, "When you look at all of the issues that manufacturers face with respect to supply chains, like third-party outsourced operations, they're taking what would normally be an internal operation or process and opening the four walls. But the responsibility is still within—they just don't have the visibility they once had." The only way to protect outside a company's "chain of custody" is to require greater information about the products. This requires three things:

1. More detailed information about movement of products through the supply chain
2. Real-time information
3. Tight integration with business systems (e.g., ERP, business intelligence, and CRM) to take full advantage of this information across the business

Real-time information is always a challenge, but it is the third parameter—integration and business utility—where numbering usually falls down. Numbering suffers from being siloed: it's useful to individuals or departments (such as file folder naming conventions), but requires a full infrastructure rollout to see any benefits.

## The EPC Numbering System

The Electronic Product Code (EPC) is an identification scheme for universally identifying physical objects. It was originally developed to be the evolution of the Universal Product Code which is found on the bar-code labels of billions of products in stores all over the world.

Although the EPC numbering system was designed to be used on RFID tags, it can be used with bar codes or even stand alone. The EPC identifier is actually a meta-coding scheme designed to support the needs of a variety of industries through a combination of new and existing coding schemes. These coding schemes are referred to as *domain identifiers*, to indicate that they provide object identification within certain domains such as a particular industry or group of industries. These coding schemes include:

- General Identifier (GID), a serialized version of the Global Trade Item Number (GTIN)
- Serial Shipping Container Code (SSCC)
- Global Location Number (GLN)
- Global Returnable Asset Identifier (GRAI)
- Global Individual Asset Identifier (GIAI)
- U.S. Department of Defense (DoD) Construct

The EPC tag data standard also defines one general identifier (GID-96) which is independent of any known, existing specifications or identity schemes, and which is composed of three fields:

- *General manager number*, that identifies the company or organization that is responsible for maintaining the next two numbers
- *Object class* (which can indicate a particular product, e.g., a 17-inch bicycle tire, a 100-count bottle of 20-mg Prozac tablets)
- *Serial number* (indicating a specific bicycle tire, bottle of Prozac, etc.)

EPCglobal and its volunteer working groups greatly advanced the IOT, in which every object may, through an affixed RFID tag, contain a unique identifier.

It is important to note that these are numeric coding schemes made up of a string of binary digits; an EPC-encoded RFID tag by definition does not hold information regarding an object's provenance, owner, and so on. Likewise, an RFID-enabled passport does not contain the holder's name, birthdate, or personal information, any more than an E-ZPass transponder or credit card holds any financial information of the holder. All of that information is typically stored in highly secure databases as it should be.

Yet, if all of this information is simply kept locked in a database, there's not a lot of business value to be gained. Therefore, the EPC Information Services (EPCIS) standard was developed to enable disparate applications to securely share EPC and related data both within and across enterprises.

The Object Name Service (ONS), which acts as an Internet domain name server, returns a list of network-accessible servers that have information about the EPC being looked up. It's important to understand that the ONS does not contain any data about the EPC.

The combination of these standards can provide an enormous business benefit to trading partners. For example, using the EPC data on an RFID tag, one could determine that this RFID tag is on a box of widgets manufactured by Acme Corp. and it was last seen in the Acme distribution center (using a Global Location Number) yesterday at 2:31 (by the date/time). It may even provide the business process step, such as "moved to sales floor."

## Numbering versus Serialization

Serialization—1, 2, 3, 4—is different from numbering. Sometimes, for safety and security, a numbering convention is purposely random. Other times, also for safety and security, a number must be serialized.

One example of a nonsequential numbering convention is our social security numbers: identical twins who apply for SSNs at the same time will have two very dissimilar numbers. MasterCard numbers are purposely random, which protects MasterCard holders from error; it does so because the 20-digit MasterCard numbering convention creates $10^{20}$ possible card numbers (in theory, enough numbers for every human being who has lived, lives now, or ever will live). If at any one time 150 million of these numbers are in use and a clerk makes an error in entering a number, the likelihood is only $150,000,000/10^{20}$ (effectively zero) that the number actually belongs to a cardholder. Here, randomness provides security.[31]

The standard Global Trade Item Number we described in the previous section does not uniquely identify a single physical object, only a particular class of object, such as a particular kind of product or SKU, for example, bottles of over-the-counter antihistamines. The Serialized Global Trade Item Number (SGTIN) allows differentiation between two like objects: one bottle of antihistamines versus the identical bottle to its right on a shelf.

Why is that useful? Because if bottle XXX-XXXX-1 is tainted, subject to recall, or expired, then XXX-XXXX-2 may also be located and removed. Or, a bottle that lacks such a number is easily identifiable as counterfeit.

This is the idea behind the California ePedigree regulations on pharmaceuticals, which mandate "a unique identifier (serialization number) placed on the smallest container saleable to a pharmacy, by the pharmaceutical manufacturer."[32] The idea is traceability of that container (be it a bottle, ampule, tube, etc.) from the manufacturing floor to "each successive owner," meaning the purchaser or patient. This is chiefly an anti-counterfeiting measure. According to the State of California, counterfeit prescription drugs account for as high as 30 percent of the supply in some countries. (Think of all those spam messages offering Viagra and "prescription meds at over-the-counter prices!")

The World Health Organization (WHO) estimates that in developed countries such as the United States, counterfeit drugs are less than 1 percent of the market; but WHO estimates that 3.4 billion prescriptions were dispensed in the United States in 2006, meaning 34 million of them were filled with counterfeit medicine. In 2006 and 2007, those developed countries, including the United States, France, and Belgium, were plagued with counterfeit oseltamivir, a flu medication; the counterfeit contained lactose and vitamin C, and no active retroviral.[33] Patients were thus fighting the flu with the active ingredients of milk and orange juice, in minute dosages.

Pharma is not the only place that serialization is vital; it similarly can distinguish between cows in a herd in the case of tainted beef, cases of bad lettuce, electronics with a particular defect, and so on.

## The Challenges

Electronic pedigrees are not easy. ePedigree will require the manufacturers to implement an "interoperable electronic system," an electronic track-and-trace system, be it using bar code, 2D bar code, RFID, and so on. That's a lot to ask, and this legislation, which was written in 2006 with a due date of 2009, has been pushed off to 2015. Why, if ePedigree seems so vital?

Because, as President and CEO Kathleen Jaeger of the Generic Pharmaceutical Association (GPhA) described, the delay "will help manufacturers to determine the most cost-effective and efficient approach to establishing an electronic track and trace system." The National Community Pharmacists Association (NCPA) was more blunt, describing a proposed 2011 deadline as "a logistical and financial nightmare for all of the affected parties."[34]

An ePedigree includes an enormous amount of information, including:

- Source of a dangerous drug, including the name and federal manufacturer's registration number
- Trade or generic name of the drug, the quantity, dosage form, and strength
- Date of the transaction
- Sales invoice number
- Container size, number of containers, expiration dates, and lot numbers
- Shipping information including the name and address of each person certifying delivery or receipt of the dangerous drug

And so on.[35]

Still, it's purposeless to fight. Whatever the objections, ePedigree is on the books in California. Any pharma company outside of California risks losing that colossal market through noncompliance. And 35 other states have similar legislation in the works.

## The Answer: A Purposeful Lean Rollout

Just as the Walmart and DoD mandates once did, ePedigree will drive a step change in supply chain utility. But the slap-and-ship minimal compliance with which a Walmart supplier could get by is not enough; manufacturers will be tasked to provide a chain-of-custody record, on demand.

Business has been down this road before, and recognizes the inevitable. Just as consumer packaged-goods manufacturers learned and evolved as a result of the Walmart mandates, pharmaceutical companies accept that they are better off in attempting to realize some business value from ePedigree mandates than in killing themselves trying to resist the mandates.

In a presentation at RFID World 2008 about ePedigree, Andre Pino of Acsis described the ROI that is possible with sequencing and numbering, which includes

- Improved receiving, packing, picking, shipping
- Automated, real-time inventory visibility (on the level that American Apparel retail locations enjoy)
- Optimal inventory levels and higher turns
- Reduced loss and theft
- Downstream visibility
- Faster order to cash (chiefly through electronic proof of delivery)
- Data integrity for audit trails
- Data for timely analysis for corrective measures and action

In short, sequencing and numbering embody all of the ideals for RFID to which industry has striven since the Walmart and DoD mandates.

Where Lean Wireless expertise is required is in making this transition smoother than the Walmart/DoD transformations, and in leveraging the existing equipment and conventions. EPCglobal's Health and Life Sciences Group ratified a sequencing standard in 2007 in answer to ePedigree, and the standard calls for XML-based codes, something Web-based and searchable with existing tools, ones that pharma manufacturers likely have in place.

## EASE THE DISEASE OF WIRELESS OWNERSHIP

In their book *Lean Solutions*, Lean authors Jim Womack and Dan Jones made an excellent point about owning an automobile: a car enables you to take jobs beyond walking distance, socialize outside your neighborhood, buy more and heavier goods, and see and do more. But these are only the positive experiences of ownership; it also entails cost and labor, of

- Registration
- Repair, overhaul, and maintenance
- Excise taxes
- Learning the technology (to drive)
- Security
- Parking
- Insurance
- Gasoline
- Time and effort of registration, driver's education, researching the purchase, driving to and from maintenance appointments

All told, an automobile costs its owner far more than the $30,000 he pays for it. It saves time and labor, but generates new costs in time and labor. However, these costs (and inconveniences) can be minimized; and dealerships, insurance companies, and repair shops compete by driving down those costs and improving service for a better ownership experience.[36] Is there a way to create a better, lower-cost experience for a company migrating to Wireless?

There has to be. Modern enterprises that want to migrate to Wireless will likely have a low threshold of pain. Womack and Jones observed that "… the software world, with its allied hardware makers, has long embraced the notion that consumers will accept products that frequently fail to work, in return for a steady stream of new capabilities and cutting edge performance."[37]

Wireless first proceeded along that same business model, but no longer. Enough is enough, cried Walmart suppliers, and those half-million- to multimillion-dollar early RFID implementations came to a halt. At present, companies that implement Wireless demand products that work, have a proof of scalability, and achieve 99.5+ percent accuracy. Increasingly, they are also demanding the subscription business model which they have come to enjoy in software as a service (SaaS). They want to use Wireless technology, but don't necessarily want to own it, maintain it, or pay extra for upgrades.

The richly RFID-enabled supply chain has yet to materialize simply because all companies do not adopt technology at the same pace. An RFID-enabled process at one company may have enormous benefits for that company, yet not suit their suppliers or trading partners. One common barrier to the adoption of EPC/RFID technology for Walmart suppliers is the lack of warehouse management systems in their facilities; they still perform inventory control using paper and clipboard. They recognize the benefits of Wireless, yet refuse to suffer any cost or disruption to implement it.

Here again, the Lean approach is necessary to make the connection between technology and the process improvements and set realistic expectations for the technology. Wireless can ultimately save enormous amounts of busywork in maintenance, production, and other processes, while generating enormous amounts of busywork for the IT department. In essence, rather than outsource the pain of ownership, poorly implemented Wireless will insource the pain to the IT department.

## The Answer: Find the Right Tool for the Right Job

The Wireless answer to most business challenges is usually the simplest, most effective, and most cost-effective answer possible. But sometimes Wireless is not the answer at all. Remember that earlier in this chapter, Entigral CEO L. Allen Bennett concluded that a prison did not need a real-time location system to track tools; it needed a peg board. But the prison did require RTLS to track keys, which were breaking off in locks; prisoners who had little else to do could look for key tips and fashion new keys in the prison machine shop. That was a better use of RFID technology, which Bennett did recommend.

RTLS was also the answer to preventing damage to the space shuttle during construction. The inventory challenge was to remove any tools that went into the shuttle. Technicians work on suspended walkways, which keeps the pressure off the space shuttle components. If a technician left a tool on the walkway and the walkway tilted, the tool could damage a component. The answer was a simple inventory system that ensured that any tool that went into the shuttle was taken out again, a cost-effective and Wireless answer to a specific business challenge.

The Florida Court System implemented a simple passive RFID system to track the last known whereabouts of any case file, versus a more complex system that could trace every file on a real-time grid, a difference of roughly $.50 and $50 per file. This is an example of a well-done cost/benefit analysis. For a $.50 surcharge per case, the Florida Court System can locate a missing file in minutes. For an additional $49.50, it could find those files instantaneously, but the $.50 solution did the job.

The technology exists now to plug practically any item into the IOT, to track it at surprisingly long distances, even to track it across borders.

Omni-ID of Foster City, California, in particular, has set new performance records with passive RFID tags (ones without batteries) by developing an energy gathering structure superior to traditional metallic-based antennas. These tags work both on, off, and near metal as well as on or even submersed in liquids. The company has partnered with IBM to offer a co-branded Data Center Resource Management (DCRM) solution, which affixes those RFID tags to IT assets. This enables the users to track tens of thousands of assets (like hard drives) across a company—for example, to conduct inventory, or ensure destruction (in the case of high-security government assets).

On the more "butch" side, the company also offers rugged encased tags that keep track of 17,000 tools and pieces of equipment—and 300 miners with helmet-mounted tags—at the Billiton Mitsubishi Alliance (BMA), Australia's largest coal miner and exporter. Finally, in 2009 Omni-ID unveiled its Ultra tag, which the authors tested in 2009, and which has a read range of up to 135 feet; we had never seen range like that from a passive tag, and few tags as rugged. This has opened up tremendous opportunities in manufacturing, logistics and defense, where asset tracking was impractical using battery powered tags with a limited life. Perhaps most importantly, the Ultra is balanced to perform well across geographies. As a passive tag, it does not transmit any frequency; it reacts to the frequencies that read it. Countries worldwide have varying standards, but the Omni-ID equipment can cross borders, so is uniquely suited for military and logistics.

## Focus upon Essential Data

One of the hesitations that Walmart suppliers had, and other companies have in plugging in to the Internet of Things, is that new equipment and the reams of data it captures would require server upgrades. Ultimately, less than a third of Walmart suppliers who implemented RFID were required to upgrade their servers, and a good many of them may have wasted their money. That is to say, they may have captured far more data than was necessary.

"There's a lot of misinformation about a 'sea of data,'" produced by RFID, Forrester Research Senior Analyst Christine Overby told the authors in a 2005 interview. "The only data [users] really need is that business insight, the needles in the haystack. For example, RFID will give you a lot of shipping and receiving data. But you don't need to worry about every single electronic product code you send or receive, only the ones that go late to your customer, or go out of stock."

And some out-of-stocks simply matter more than others, for example, on consumer electronics with their high margin, versus a commodity product such as a bag of loose candy. "So you won't have to monitor this ocean of data" observed Overby, "*just the data that matters.*"* (Overby has since become a vice president and research director at Forrester.)

With Wireless technology, a company has the ability to collect an ocean of data among:

- Equipment condition and informatics
- Asset location, movement, and utilization
- Inventory levels
- File location, copying, and transfers
- Internet access and usage
- Thousands of other bits of data

What companies doing business with Walmart in 2004 discovered was that they did not require a bank of servers; rather, they required data filtering and strong middleware applications. And they required expertise in identifying key information, chiefly business events, such as the shipment of a product or an impending stock-out.

## Practice Agnosticism—It Is Lean

The word "agnostic" originally meant being unwilling to commit to belief in a deity, but has come to be a catch-all term for an unwillingness to

---

* Italics ours.

commit, be it to a political party, employer, and so on. In a person, that might seem wishy-washy, but in technology, it is ideal. "Agnostic technology" as it is commonly called, integrates with any platform or enterprise software, and conforms to numerous standards, as does Wireless.

Office buildings, manufacturing plants, distribution centers, commercial environments, and college campuses use a standard called 802.11. The 802.11 Wi-Fi (or "wireless broadband") is largely "plug and play." Any computer or device equipped with 802.11 wireless capacity can connect to the network. Moreover, a wireless-enabled broadband modem is typically connected to the Internet, or an enterprise's intranet; this allows a company to connect any two nodes on its Wireless nerve path, be they equipment or people, as we saw in the previous section.

Vendors are now creating platforms that allow interoperability between all devices in an agnostic environment, which enables end users to select the "best of breed" wireless sensors and devices, regardless of the standard used.

## Devices, Processes, and People Become "Nodes" on the Nerve Path

Wi-Fi is largely plug-and-play. As stated above, any computer or device equipped with 802.11 wireless capacity can connect to the nerve path, and each becomes a node. This includes:

- Mobile computers such as laptops
- Walkie-talkies
- Programmable logic controller (PLC) devices (used for automation)
- PDAs
- Wireless sensors

Moreover, applications and processes connect to the nerve path, and become more nodes. This includes:

- Telephony, using VoIP
- RTLS
- Inventory management systems
- Logistical processes, such as advanced shipping notices
- Line-of-business applications such as computer maintenance

Finally, because the nerve path is Web based (using the Internet or a company's intranet), a company can connect almost any two nodes over the nerve path and thus connect any process to any device and vice versa. Some examples include:

- Remote sensor alerts directed to a PDA
- Security alerts sent to smartphones
- Advanced shipping notices triggered by an RFID portal on the outbound dock door
- An order generated for medical supplies by an RFID-enabled smart shelf
- A replenishment order generated by an eKanban system, which uses RFID, bar code, and other inputs

Wi-Fi is ideal for more than just mobile computing and connecting nodes. It is a global standard defined by the Institute of Electrical and Electronics Engineers (or IEEE) with nearly 375,000 members in more than 150 countries.[38] It is supported with technology from such powerhouses as IBM, Texas Instruments, HP, Motorola, Nortel, Intel, Microsoft, Ekahau, and Siemens.[39] And it is proven at some very exacting locations, such as the Disney theme parks in tracking children, FedEx in tracking parcels, at Toyota, Walmart, the DoD, and on university campuses and at Starbucks.[40]

Moreover, Wi-Fi benefits from the "consumer effect." Consumer-grade wireless networking—a Wi-Fi Internet connection—is available in homes for as little as $30, and an enterprise-grade Wi-Fi router can cost as little as $200. Similarly, consumer-grade GPS widened its appeal and fostered lower costs, and enterprises use it routinely in logistics and in yard management.[41]

At present, Wi-Fi is the most mature of wireless networks. The other wireless technologies, such as UWB, IR, ZigBee, and the like, all of which have practical utilities, simply do not have the footprint of technology providers and adopters.[42]

For an enterprise to effectively use and manage its devices and device data, it will require a hardware abstraction layer, which provides a common interface for the disparate devices. Additionally, given that not all devices have the ability to communicate with more than one system at a time, a control layer is also required to coordinate conversations from multiple systems (e.g., to record temperature data in one system and send alerts to another system).

A network with this level of sophistication requires a platform that facilitates the creation of an intelligent sensor network or ISN.

## If Only It Were That Easy: Connecting the Edge and the Back End

Simply put, old-model networks cannot handle smart assets and smart edge devices. They were sufficient when the only hardware nodes we had were printers; but let's take a look at what is realistic now (see Figure 3.1).

There are five layers to an old-model network. Starting from the bottom, the layers are:

- The *edge* (such as a plant floor), where both workers and devices (such as computers and sensors) are
- The *extended network*, where WLANs, RTLS, and the like reside
- *Edge controllers* and *servers*, which manage the devices on the edge
- A *middleware layer*, where intelligence is added
- The *back end*, where the enterprise system (such as SAP or Oracle) connects it all—if indeed

Thus, one does not simply "plug in" a Wired or Wireless sensor or a new business process to the backbone. Old networks, which had difficulty plugging in software applications such as customer relationship management (CRM), are not at all equipped for dozens, let alone hundreds of new processes and nodes. To do so requires integration expertise—which costs time and money—and so the dreaded "silos of information" exist, information that is not visible from the "shop floor to the top floor," or between departments.

And back-end integration, long thought the ideal, is simply overrated. Somewhere along the line, everyone accepted that every byte of information, every process, every application, every indicator, must be integrated into the enterprise backbone. Or, that data must connect seamlessly from the shop floor to the top floor, so that even C-level executives have visibility into "the edge."

That requirement is unnecessary, and likely kills a lot of useful projects. According to industry think tank Aberdeen Group, data integration represents over 60 percent of the cost of any automated data capture project.[43]

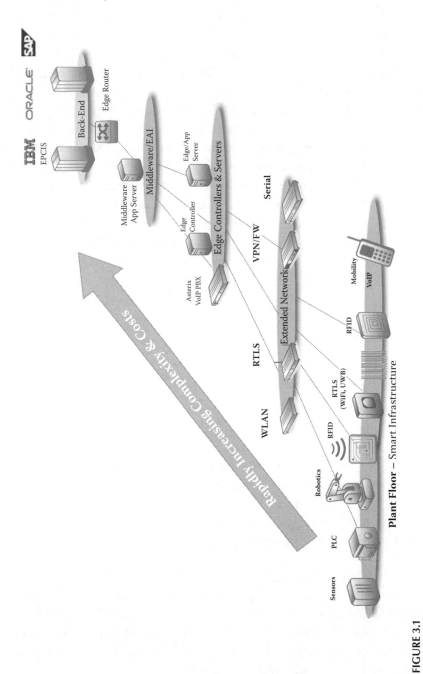

**FIGURE 3.1**

Integration on an old-model backbone. Bleeding heartbreak. (Courtesy Omnitrol Networks Inc.)

And that diverts resources away from production goals. "The value proposition of shop-floor-to-top-floor integration is often not well thought out," said CEO Raj Saksena of Omnitrol Networks. "Information on the shop floor is usually *for* the shop floor," and of no use to anyone at the back end. Certainly, the C-level executive needs intelligence into shop floor operations, but just how much intelligence? The answer is, which intelligence can that executive use?

Let's undertake a theoretical exercise in purposeless integration. United Parcel Service (UPS) has speedometers on the dashboards of its brown trucks. But maybe there's a more efficient way to process the information of speed. We have the capability to equip vehicles with remote speedometers, which the drivers never see. These could be represented on a sophisticated "executive dashboard" that allows a regional manager—no, wait, let's go right to the top floor—that allows UPS's VP of Operations for the Americas to monitor driver speed, and manage that speed "by exception," phoning any driver that he believes is driving too fast for safety, or too slow for efficiency. Such connectivity would provide seamless visibility from the edge to management, from the shop floor to the top floor—and it is absolutely possible, with remote sensing and cellular signals! It is also absolutely purposeless.

## Question Integration, Always

When faced with an integration decision, you need to answer four primary questions:

1. Is the intelligence truly necessary, or value-adding? (Or otherwise required, for example, in ePedigree requirements?)
2. Who specifically requires the intelligence?
3. Is back-end integration absolutely necessary?
4. Is historical data absolutely necessary?

Chances are, a good many classes of information are like vehicle speed: necessary in the moment, to a specific individual. Retaining and monitoring the information serves no purpose to anyone but that individual. Passing intelligence through the ERP or warehouse management system (WMS) may simply delay it, obscure it from those who need it, or worse, discourage collecting that intelligence at all.

The answer is not in surrendering that integration and visibility, but questioning its value (always) and vastly simplifying it. The answer is a new, and simpler, infrastructure: the smart edge infrastructure.

## Smart Edge Infrastructures

The smart edge infrastructure model that Omnitrol and some rivals offer combines the middle three layers (of controller/extended network/middleware) into a single hardware appliance; this appliance is agnostic in that it can accept any Wireless device of any make. In essence, the devices integrate with the appliance, which in turn, integrates with the back end (see Figure 3.2).

"Our intention was to give the shop floor what it needs, quickly as possible; real-time alerts on dashboards, using the existing infrastructure, and allowing them to get more effective and efficient," said Omnitrol CEO Raj Saksena. "Sure, we integrate to the back end, but we don't start at the back end and move down; we start on the shop floor," or the edge, "and move up." The changes are at the edge, not on the enterprise backbone, which is more costly and difficult.

The value propositions of such a model are considerable. Omnitrol claims:

- More than 10× reduction in deployment cycle time over competitive enterprise solutions
- A 5:1 total cost advantage over back-end solutions
- Linear cost scalability
- Multisite scalability
- More immediate ROI by eliminating back-end re-engineering, complex middleware, and field support
- A single integration point for mobile handheld business services, bar code, RFID, Wi-Fi, UWB, PLC, and sensors
- Hot-install—the ability to "flip the switch" and plug into the existing work flow—which eliminates downtime

## Simplified, Lower-Cost Visibility

The challenge of plugging Wireless into the classic enterprise backbone lies in multiple layers of smart devices, silos of information, and complex intelligence layers and middleware. Omnitrol provides an architectural

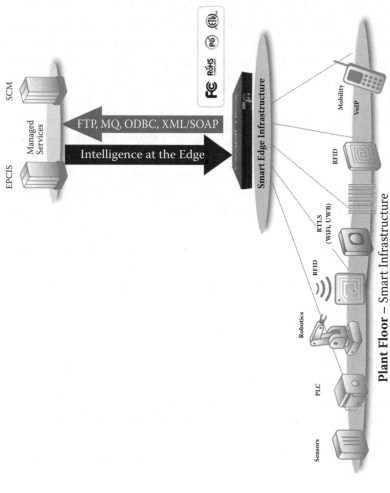

**FIGURE 3.2**

The intelligence and power sit at the edge. (Courtesy Omnitrol Networks Inc.)

solution with a new generation infrastructure, with applications that provide full visibility without having to integrate the devices through those interim layers to the ERP back end (which can take weeks).

"That was the value proposition demanded by the market," said Saksena. Companies required a network that would integrate:

- Newer technologies including RTLS, RFID, and Wi-Fi
- Legacy technologies such as robotics and PC controllers
- New mobile computers and devices (chiefly notebook computers, tablet computers, and PDAs)

"This new world of sensors and value-add automation led us to believe the way to do this was to create a single versatile programmable network appliance," Saksena continued, one which connects any device to the network, allowing instant visibility and control. With such an appliance, a PLC controller or RFID reader plugs into the infrastructure in minutes, versus weeks.

---

## MANAGE THE TROUBLESOME SOCIAL/ ENTERPRISE INTERFACE

> I'm mobile, and wherever I go, I'm connected. I work hard and I play hard, and sometimes I like to play a little at work, just to keep it alive.
>
> **Actual and unabashed quote from the blog employee2dot0.com**

Because the Wireless revolution is to some degree consumer and culturally driven, it is impossible to separate business from consumer utility. Work feels like play because the tools are the same. Millennials and digital natives do not necessarily lack scruples, but, they lack clear boundaries, or a clear understanding of the challenges they present their employers.

This attitude creates real consequences. The Obama–Biden transition team famously used a seven-page questionnaire to vet prospective appointees. In addition to the usual stuff (e.g., disclosing financial records and college transcripts), the questionnaire required applicants to detail any "aliases or handles" used on the Internet, and to reveal any e-mail or other electronic communication that "could be a possible source of embarrassment to you, your family, or the President-elect if it were made public"; applicants were not safe from any raunchy video or ethnic joke mailed a decade ago. David Gergen, the sometimes-Republican, sometimes-Democratic adviser to both George H. W. Bush and Bill Clinton, described the questionnaire as "invasive."

Perhaps. Within two weeks of his inauguration, Obama's chief speechwriter, 27-year-old Jon Favreau, became infamous for a photo of himself at a Chicago party, groping the breast on a cardboard cut-out of Hillary Clinton, supposedly snapped with a digital camera and uploaded to Favreau's Facebook page (see Figure 3.3). (Twenty-seven is younger than it used to be.) No low-level staffer, Favreau is credited with penning Obama's catchphrase "Yes, We Can," and also penned Obama's acceptance speeches, for both the Democratic and presidential nominations. Not even beer-fueled late night foolishness is private.

Wireless Communication has removed a layer of judgment and decision making. This creates more risk than waste, but nonetheless, the

**FIGURE 3.3**
Spontaneous humor in the Internet age. Beer + Wireless + instant upload capability – judgment = embarrassment.

consequences can be dire. As of this writing, Google is testing a feature called "Mail Goggles," which puts five simple math problems between writing and sending an e-mail, using Gmail. Someone without the presence of mind to solve five equations on the order of "69 – 30" and "27 + 13" is likely writing something moronic. Mail Goggles is designed as an option, which an individual can set to police him- or herself during vulnerable times, such as late on a weekend night, after more than a few beers. There is yet to be an enterprise version of Mail Goggles, but likely, there will be.

## The Real Costs of Play at Work

If in the 1970s your father worked in an office (chances are, in the '70s, your mother didn't work), he likely brought home a photocopy of "The Last Great Act of Defiance" (see Figure 3.4). This crudely sketched humor was common back then, a harmless little diversion to pass the time. Let's do some simple math to determine the cost to business of this li'l bit of levity, assuming 50 pass-alongs, from your dad to a few friends, and a few friends of theirs:

$$50 \times \$.005 \text{ cents/page} + \text{lost productivity} + \text{wear}$$
$$\text{and tear to photocopier} = <\$5$$

Not so bad.

Jump ahead 30 years to 2005, when the Dead Santa pass-along (Figure 3.5) emerged, an electronic file presumably put together with some application like Photoshop. Unlike "Defiance," Dead Santa suffers no deterioration through generations of copies. Let's assume 50 pass-alongs again:

$$(1 \text{ MB [inbox of originator]} + 1 \text{ MB [outbox of originator]})$$
$$\times 50 \text{ pass-alongs} = 100 \text{ MB server space, nominal}$$

Would that those were the only costs: if each copy of Dead Santa were infected with malware, and each terminal must be reimaged, and we assume (nominally) 50 terminals at a cost of (nominally) $400 in lost productivity and IT labor, then Dead Santa has cost a business $20,000.

But as the quote from Employee 2.0 ("I like to play a little at work, just to keep it alive") embodies, the Millennial feels entitled to the latest technology, and the latitude to use business systems and equipment for play. They

**FIGURE 3.4**
Pass-along office humor, circa 1975. A classic, and you couldn't attach a virus to it.

feel entitled because they perceive themselves to work so damned hard; presumably, much harder than their parents or grandparents who had no such sense of entitlement.

Accenture Technology Consulting surveyed 400 of what the firm called "Millennial generation students and employees (those aged 14 to 27)"

**FIGURE 3.5**
Pass-along office humor, circa 2005.

and found that the Millennials "… Expect to use their own technology and mobile devices for work and are increasingly choosing their place of employment based on how accommodating companies are to their personal technology preferences."[44]

In addition, 60 percent of Millennials were either unaware of their companies' IT policies or simply disinclined to follow them. Six percent were completely aware of their companies' policies but actively chose to ignore them. "The message from Millennials is clear," said Gary Curtis, managing director of Accenture. "To lure them into the workplace, prospective employers must provide state-of-the-art technologies … And if their employers don't support their preferred technologies, Millennials will acquire and use them anyway, be they smartphones, public Web sites, freeware and software applications or the like."[45]

The challenge is that if an IT department has no control over the equipment or applications that employees use, then it cannot safeguard enterprise information.

## So Long, Lunchpail

All that aside, the Millennials have considerable strengths. Theirs are the skills an enterprise requires to compete.

In the 1960s United States, an estimated one-third of the workforce was on the factory floor, versus less than 17 percent in 2006.[46] "Eventually, manual labor will disappear in the factory and in the home. A world of robots is our future," opined Professor Sohail Inayatullah, a political scientist/futurist associated with Tamkang University in Taiwan. "We will be designers, engaged in serving others," in services, finance, and conflict resolution. Blue-collar factory jobs will decline sharply in Organization for Economic Co-operation and Development (OECD) nations by 2020, but not globally, just as farming and agricultural labor has declined in once-strong agricultural centers.[47] In the United States, the percentage of laborers who are unionized is under 12.5 percent, down from a third in the 1950s.[48] In short, the nature of work in the United States is changing.

Henry Jenkins, author of the books *Convergence Culture*[49] and *Textual Poachers*, among others, has identified a new set of core social skills and cultural competencies that young people must acquire and schools must embrace, "If we're going to fully prepare every kid in America to be part of the participatory cultures" which America has designed for itself. Four key skills are:

- *Collective Intelligence*, or the ability to pool knowledge and compare notes with others toward a common goal
- *Judgment*, or the ability to evaluate the reliability and credibility of different information sources (which, on the Internet, is subject to practically no vetting or peer review)
- *Networking*, or the ability to search for, synthesize, and disseminate information
- *Negotiation*, or the ability to transcend communities and perspectives and accept alternate norms

Thus, the performance evaluation in 2015 will likely include numerical measures, such as a networking acumen index or virtual collaboration index, a measure of if an employee "plays well with others online."

Chances are, that employee will play well indeed. The Millennials are highly trained digital natives as opposed to the "digital immigrants" (older workers) who must overcome old paradigms and bolster old skills sets.[50] The digital natives have agile minds and are used to virtually no support. "Give a new game to a digital native [and] it takes him three to five minutes to intuit everything he needs to know," futurist Glenn Hiemstra told *BusinessWeek*.[51] "If you're thinking about how you'll use IT in your businesses 20 years from now, don't think about yourselves—think about the digital natives. They will be the work force, and they'll take it beyond what we imagine. This is the way the world has always been." Modern operations and IT officers can plot their companies' strategic directions, and HR policies, and IT priorities, by observing their kids; their use of technology will become the norm when they enter the workforce: especially in 15 or 20 years when they are the company officers. And so the enterprise interface of 2015 will owe as much to *The World of WarCraft* as it will to Microsoft Windows.

*WarCraft* is a massive, multiplayer online role-playing game, or MMORPG. Companies such as IBM and Wells Fargo are not blind to the downfalls of MMORPGs (hundreds of hours of wasted time), but have accepted its strengths as well, such as the ability to assemble remote teams to a specific task; it's a leap between joining together to bash a troll in *WarCraft* and designing a medical device in industry, but not much of one. The transferrable skills are the ability to identify and create a team based on specific skills regardless of geography, and the ability to disband a team without hurt feelings, layoffs, or relocation.

## Millennial Learning Is Lean

In an ominously entitled report called "The Ill-Prepared U.S. Workforce," a consortium of human-resource heavyweights* found some good news about Millennials. True, they require some training in workforce preparedness; but they like to learn new skills, and take the initiative to learn, provided you give them the tools they like. And, those tools are informal and cost effective.

The majority of the 211 companies surveyed use "informal training methods," ones outside of classroom or training sessions. Almost three quarters of those companies (70.8 percent) cited intranet access as an effective training tool, 63.1 percent cited e-mail for sharing knowledge, and voluntary informal mentoring was third at 60.0 percent.[52]

Good, as far as it goes. But only 18.5 percent of employer respondents cited online groups, and 10.8 percent cited social networks (e.g., Facebook,

---

* The Conference Board, the Society for Human Resource Management (SHRM), the American Society for Training & Development (ASTD), and Corporate Voices for Working Families (CVWF).

MySpace, LinkedIn),[53] tools with which the Millennial workforce is eminently well trained, and likes to use, and which are cost effective and spontaneous, say the companies surveyed. Two thirds of respondents found that informal training occurs naturally, and 63.1 percent find it cost effective.[54] They require cost-effective training, as U.S. companies slashed formal training budgets by 11 percent in 2008.[55]

A company that wishes to create a truly "learning organization," one that both learns and teaches itself, must make use of those tools.

## Millennials Question Authority and Policy and Are Not above Defying Them

The downside of those cost-effective tools is that Wireless Communication and Web 2.0 so empower Millennials that they have not only a defiance for limitations upon their usage, but a grandiosity about their own talents and work ethics.

As Inayatullah put it, Millennials believe that "novelty and innovation does not come from above—the Great Leader hypothesis." Rather, the purpose of the leader is to create capacity of those in the business to innovate and create, so that they do not feel created upon.[56] Millennials seek novelty, and trust themselves to innovate.

They should, to a degree. Youthful innovation *is* rewarded on Web 2.0. Facebook, Twitter, and CraigsList are grand-scale Web 2.0 successes started by young entrepreneurs. But grandiosity without talent is also rewarded. Observe the long list of "Webutantes" like Lisa Nova, a breathtakingly talentless Internet sensation who produces short "comedy" videos on YouTube; she has more than 160,000 subscribers (and counting), 10 million views (and counting), and 11,000 Twitter subscribers (and counting). She was too popular to ignore and parlayed her Web fame into an appearance on the variety show MadTV, but the producers declined to sign her on as a feature player. Similarly, with digital film cameras costing less than a month's salary, movie making has become the new community theatre, and the Telluride and Sundance Film Festivals are choked with amateur-quality entries.

## Tales beyond 2000: The End of Secrets

Modern business is finding that secrets are harder to keep in a world where its new workers have little expectation of privacy; are equipped with cellular phones that can record every sound, sight, and deed; and can use flash drives, e-mail, and Wireless computers that can send megabytes of company data through and around any firewall.

"That broad set of technologies for recording and sharing one's observations—these technologies undermine secrecy, whether for good or ill," said futurist Jamais Cascio. He coined the term "Participatory Panopticon," which refers to the willing and ongoing recording of one's thoughts and deeds, be it in public places or private. The Panopticon was a prison design, conceived by philosopher Jeremy Bentham in the eighteenth century (see Figure 3.6).

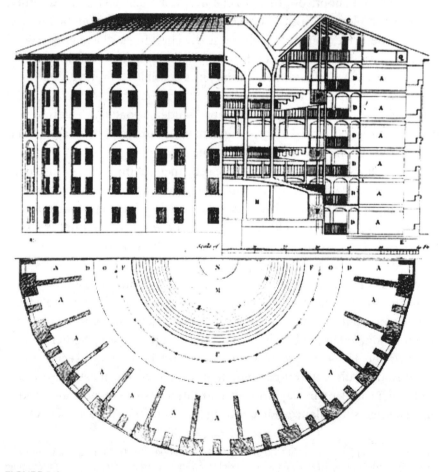

**FIGURE 3.6**
Jeremy Bentham's Panopticon design. No privacy for prisoners (but the prisoners wouldn't know that).

Its design isolated prisoners from one another, but made all prisoners visible to observers at all times (and unbeknownst to the prisoners). They would get away with nothing. The Panopticon conveyed what an architect called a "sentiment of invisible omniscience."[57] Bentham described his design as "a new mode of obtaining power of mind over mind, in a quantity hitherto without example."[58] *

Likely, if you put it to young workers in those terms, they would be horrified. Yet, they go willingly into the Participatory Panopticon, sharing every thought and deed on blogs, walking by closed-circuit televisions without hesitation, purchasing cellular phones with GPS capability. In the Participatory Panopticon, the citizens monitor themselves, one another, and their employers. This is "sousveillance," surveillance from below.

Cascio describes the camera phone as the "harbinger of a massive social transformation, one already underway." Radio and television are passive reception media; with a camera phone, it is possible to capture a photo or video and e-mail it or post it to a Web page. A popular subject on YouTube is school fights and freakouts by teachers; students goad teachers into verbal, sometimes physical, violence, capture it on camera phones, and post it on YouTube before leaving the classroom.

On the plus side, Cascio expects the sentiment of invisible omniscience to stimulate a work ethic by making employees far more cautious about being caught. It will also create master criminals. "They'll pay more attention to keeping things hidden. So the ironic result of making it easier to discover the casual corruption is that it is harder to discover deeply embedded corruption. The dumb ones don't last."[59]

And so, master criminals evolve. They may write ultrasophisticated malware, such as the 2008 Conficker virus, which enables remote access by hackers. Conficker has yet to be traced to a writer (although suspicion is that it originated in the Ukraine). It grounded French Navy fighter planes, infected the U.K.'s House of Commons, and its writers execute more updates and vulnerability patches than Windows. Or, master criminals evolve who find their way around regulations, surveillance, and auditing. Neither surveillance nor auditing nor the federal Sarbanes–Oxley

---

* Bentham was a highly respected Utilitarian philosopher at the time (1748–1832), and a futurist, with forward-thinking treatises on women's rights, animal rights, divorce, and free trade among other topics. He was also eccentric about continuous monitoring and observation. Bentham had himself taxidermized so that he could attend meetings of the University of London College Council and break tie votes, posthumously.

requirements (for corporate governance and financial disclosure) headed off the banking crisis of 2008–2009. Neither did surveillance nor auditing prevent rogue trader Jérôme Kerviel from costing his employer, France's Société Générale bank, $7 billion. To pull it off, Kerviel used outdated network access, which he should have been stripped of when he became a trader, and passwords which he finagled out of other employees. By all reports, the mechanisms of oversight were in place to recognize suspicious trades, but no one at Société Générale was watching.

What business lacks are the very rules that Lean experts favor, but automated rules. Société Générale had an enterprise system that raised a red flag, but did not close off access to Jérôme Kerviel. The DoD system did not disallow copying sensitive logistics and personnel files, and in the case of the Super Bowl malware, the workplaces affected either had no rules concerning nonbusiness use of the Internet, or did not enforce them with automation.

## Viral Information for Good and for Ill

A boon and a challenge to Enterprise 2.0 is the viral or exponential nature of computer pass-alongs. To understand the challenge, consider these examples:

1. *The linear progression.* One object or bit of information passes from one individual to another. This is like the kid's game of "Hot Potato" or the water balloon toss, or lending a book.
2. *The geometric progression.* Best explained by an old Fabergé Organic Shampoo commercial from the 1970s, "I told two friends, and they told two friends, and so on, and so on, and so on."
3. *The exponential progression.* Also known as "viral" progression, as in "viral marketing." This is how the modern computer environment works. One message reaches thousands, which in turn reaches millions, and it is completely accessible at any time, from anywhere. This is because one message is easily reproduced by enthusiasts, and passed along, via e-mail, text message, Web links, and the like.

Exponential progression is behind the success of such viral marketing campaigns as Unilever's Dove Real Beauty campaign, in which supposedly ordinary women posed in snow-white skivvies. With a dedicated Web page, enthusiasts were able to send links, blog about their own curves, and so on. BMW Films was another such campaign. More than 100,000 BMW-positive films exist on YouTube, most created not by the company, but by BMW enthusiasts.

On the down side, the Super Bowl virus spread exponentially, as well, as did the fairly harmless "Dead Santa" pass-along. So too does misinformation, for example, that (former) Alaska Governor Sarah Palin did not know that Africa is a continent. This was traced to a fake conservative think tank (the official-sounding "Harding Institute for Freedom and Democracy," which was just a Web site), and a fictitious McCain adviser named "Martin Eisenstaedt." His act was good enough to fool MSNBC, which broke the story but recanted it within 48 hours.[60] But the damage to the McCain campaign was done. Misinformation spreads as fast as information; faster when it is fun.

"Information on the Internet is subject to the same rules as conversation at a bar," said Dr. George Lundberg, who edited both the *Journal of the American Medical Association* (*JAMA*) and *Medscape,* an online and well-vetted portal for physicians. But information is powerful when it is free, expedient, and credible-looking. Accessibility and expediency are the norm in Enterprise 2.0, perhaps more valuable than truth and accuracy.

Consider Wikipedia, the online user-generated encyclopedia. According to Web-traffic enumerator Alexa, Wikipedia received roughly 450 times more traffic than the online version of the *Encyclopedia Britannica* in the first quarter of 2007. It is user-generated, whereas the online *Encyclopedia Britannica* is written by scholars and subject to peer review; but *Britannica* costs $69.95 per year and Wikipedia is free.[61] Educators routinely disallow Wikipedia as a source, distrusting its free-form user-generated content. Wikipedia in January 2009 reported that Senator Ted Kennedy had died. Kennedy died eight months later, in August 2009. That same month, Wikipedia added a layer of oversight, in the form of volunteer editors, to approve edits to English-language stories about living people.

## The Answer: Airsource and Automate Decisions and Behaviors

Earlier in the chapter and in previous chapters, we described the Super Bowl XLI malware. On Friday, February 2, 2007, just before Super Bowl XLI, Websense Security Labs began getting calls from frustrated enterprise customers who found themselves unable to reach the Dolphins Stadium site at all. Simply put, employees of Websense customers did not have the choice to expose their companies; the "good choice" was automated.

Websense ThreatSeeker™ technology had discovered the malware, and disallowed access by its customers, using Websense Real-Time Security Updates™ and in time to prevent monetary loss. Websense describes what it calls essential information protection, or EIP, which is a uniquely Lean concept: it assigns granular rules to specific information, which is tagged with an "electronic fingerprint." For example, a company may specify:

- Only lead engineers may copy CAD drawings to a USB drive.
- Only marketing personnel may view pop-up ads.
- Only a specific personnel administrator may e-mail social security numbers, and never more than one in a single mail.
- Only physicians and registered nurses may read patient histories, and only physicians may enter diagnoses.
- No employee may e-mail a shipping coordinate.

This simple security protocol would have headed off the bad choice, for instance, to put defense logistics files on an MP3 player.

## The Value of Airsourced and Automated Decisions

Lean principles wisely view paperwork as muda; if filling out work travelers, trucking manifests, and expense reports were fulfilling, employees would be better at doing it. Recall our earlier point as well, that human error puts a glass ceiling on quality by introducing a consistent margin of error.

Most repetitive processes may be automated to remove that error, for example, scanning a car chassis going by in production, generating advanced shipping notices, and taking inventory. But where there is repetition, there are consistent judgments and criteria, which can be automated. Thus, there are far more opportunities for automation and airsourcing:

- *Data entry* may of course be automated, and this is where most companies start their automation.
- *Decisions* may be automated.
- *Service* may be automated, so that a customer has immediate satisfaction and resolution.

- *Observations* may be automated where objective evaluation is needed.
- *Regulatory compliance* may be automated, so that an enterprise does not risk losing data it is mandated to keep, or keep data that it is mandated to destroy.

Beyond these repetitive tasks, the behaviors and values of:

- *Judgment* may be automated.
- *Secrecy* may be automated, such that essential information does not leave an organization without authorized consent.
- *Honesty* may be automated.

At first glance, this all seems rather dehumanizing. Decisions signify empowerment, which is a Lean ideal. But a good many decisions are simply a chore. A licensed welder is empowered by deciding which tools or composites he uses on a given job, but deciding which job among several has top priority is simply a nuisance. As for paperwork surrounding a welding job, it is tempting to skip it. Thus, some decisions are simply forms of overprocessing, scut work, and bad processes.

Rather than dehumanize the workplace, automation extends the reach of human decisions that are made up front and made well. Look no further than your automobile. Recall the old rule that oil must be changed every 3,000 miles. That is not many miles; a traveling salesperson can rack up that much in under a week, a commercial trucker in two days. Somewhere in the 1990s, automotive engineers concluded that this was a waste; the quality of the oil, not the age, mandates replacement. And, the driver has a right to that knowledge. A '67 Ford Mustang had just an oil light, but the "oil life" indicator in a modern automobile measures oil acidity, indicates oil life by percentage, and tells the driver when to change oil for optimal long-term performance. Suddenly, we are changing our oil every 8,000 miles or so. (This of course has much to do with the quality of auto engines as well; still, we are relying upon measurable data and not rules of thumb.)

Business applications routinely automate decisions. Iron Mountain, with its eRecord storage, automates the decision of which records to retain, which to discard, and when. These decisions are made by cool informed heads, knowledgeable in Sarbanes–Oxley requirements for retention—perhaps by corporate counsel—and not by someone who is afraid to ditch records in these litigious times, or recklessly deciding "when in doubt, throw it out." In the case of Websense, a hospital administrator can make the decision about who has access to patient records, and can automate that decision. Those managers have extended the reach of their qualified decisions, and automated the enforcement of those decisions.

Similarly, Wireless and Web 2.0 technologies can be used to standardize processes by automating decisions wherever there is sufficient information to make such a decision in a predictable way. Some simple and real-world examples of automated decisions are:

- Reordering medical supplies and generating purchase orders automatically, using RFID-enabled smart shelves
- Shutting down rotating equipment when remote sensors detect vibration outside of tolerance
- Installing enterprise-grade security systems such as Websense on company-issued mobile computing assets including laptops and BlackBerry devices
- Deciding who shall have access to the property, then enforcing that decision through RFID-enabled keycards

The last two of these examples involve security and access. Simply put, in a business environment of extraordinary distraction, higher risk, thinner margins, and global competition, every business process must work well, and wherever possible, it must be automated.

## Not All Decisions Can Be Automated

Automated decisions are largely utilitarian, such as shutting down a heavily vibrating turbine, sometimes ethical, as in the decision to disallow broad transmission of patient records, but rarely are they life and death.

Isaac Asimov's robot novels of the twentieth century explained the dilemma well: robots serve humankind, and must be preprogrammed to do no harm to humans, and disallow harm to come to humans. But they are ill-equipped to mitigate emotional harm, or to make ethical decisions. One unfortunate android in *I, Robot* told a woman that a colleague loved her because to hear otherwise would harm her emotionally (which the lie did, ultimately). In the film version of *I, Robot*, a rescue robot made the judgment to save a man versus a little girl in a car wreck. It was a logical choice because the man's chances of survival were greater, but the ethics were arguable.

And so moving forward, only the decisions that are consistent and inarguable will be automated. Physicians, police officers, and firefighters and fighter pilots—not algorithms—will make on-the-spot, life-or-death decisions.

---

## DESIGN EFFECTIVE INTERFACES

As we said earlier in the chapter, you need to ask employees what technologies they want to use, because they will use them anyway. They must be involved in the design of Lean Wireless devices and interfaces, but they need to be guided as well. As we saw in the Super Bowl example, the convergence of business and personal use on a single device fosters a lack of security and loss of productivity.

The most enlightened companies, wrote *Information Week* editor Rob Preston, look to their employees to identify consumer functionality they can bring into the enterprise as-is (such as Google Apps and the iPhone), or buy from Web 2.0 vendors (such as Clearspace, SelectMinds, and Socialtext). Preston was writing about the Accenture survey, which found that employees demand intuitive collaboration, project management, and community tools, and that companies such as Coca-Cola, Motorola, and Procter & Gamble provide them.[62]

The technologies that Millenials "demand" make them happier and more productive, and frankly, can be much cheaper than existing tools. Some real possibilities include:

- The game interfaces that replace the Windows-style graphic user interface or GUI
- The Facebook style social network used to create knowledge bases
- Other social networks, which provide the new model for the A3 form (which employees use to suggest process improvements) and kaizen

The built-in strength of these applications is that employees like them, and were trained to use them on their own time. "We're constantly checking with each other for useful insights," said Cascio, for example, in finding a new restaurant, in choosing a political candidate, even in romantic choices. "You meet a new guy, and want to know if someone in your circle has dated him before."[63] A next-generation version of the PDA—the personal memory assistants (PMAs), and the Participatory Panopticon— will enable us to ask questions and find answers continuously.

Social network tools are what Cascio describes as the "killer app" of the Participatory Panopticon.[64] On the consumer side, when the facial recognition applications at work in military and law enforcement applications make their way into social networking—and they will—we will be able to capture a stranger's face with a camera phone or PMA, get a shortlist of possible individuals, and find out practically everything we need to know about the individual (both business and personal) that is recorded. That may sound horrifying to any reader over 30 years of age, but recall our earlier observation: Millennials grow up with little expectation of privacy, and surrender it willingly when they have it.

They also surrender human interaction, willingly. Someone who grew up making friends and enemies online does not distinguish between face-to-face and electronic interaction. Nor, in 20 years, will that individual distinguish between human and robotic interaction; the two will be barely distinguishable, and the population will interact with robots and interfaces from infancy.

One such robot in development at the Massachusetts Institute of Technology (MIT) Media Lab is The Huggable™, a robotic teddy bear for use in healthcare, education, and "social communication applications." According to the Media Lab, the Huggable is an "essential member of a triadic interaction ... not designed to replace any particular person in a social network, but rather to enhance that human social network."[65] A "triadic interaction" includes, presumably, a child, a Huggable, and a parent. But interactive technology always defies intent. Television was conceived for entertainment and information, but became an ersatz babysitter. The Internet was conceived to trade scientific information, none of which is found on Hooters.com. Users decide what is proper usage.

## The Wearable, Always-on Interface

Ultimately, between three-dimensional circuitries and thin layers, we will be able to create a PC with phenomenal processing capability, about the size of a sugar cube. Both futurists and technology providers predict wearable computers, which will evolve to include voice recognition and recognition of eye movement and gestures. Some call this the personal memory assistant, an always-on, always-connected mobile tool. We likely will not dig it from our pockets and handbags; we will wear it as glasses, or in its smallest form, an earpiece or badge (like Kirk and Spock wore).[66]

The PMA will do more than assist us: it will do things we cannot. With superior always-at-hand processing power, it will extend users' ability to categorize, collate, prepare, and search documents. We will talk to it as a thinking machine, asking, for example, "Find all documents after January 1 this year with the word strings, 'Tort costs' and 'appeal.'" Or the computer itself will categorize and collate documents, based on our historic usage.

Jeff Jacobsen of Kopin, Inc. meets us at the Industry Wizards office outside of San Francisco. He hands over a small black box, about the size of two ring-boxes stuck end on end. It looks like the single-slide viewers that photographers used to use, but instead of a vacation picture, you see a Windows interface.

"Mail," said Jacobsen, and "Number Three." An e-mail screen opens, and the third item on the list opens. The device has no keyboard, but he can record and send a voice reply, or, allow the voice recognition capability to turn the recording into text and reply with an e-mail. We observe that the SkyMall catalogue offers several liquid-crystal display glasses-shaped interfaces for watching films, which simulate the wide-screen TV experience.

"We made those," said Jacobsen of the display. Kopin introduced its CyberDisplay product line in 1997, and has shipped more than 30 million units to customers including JVC, Kodak, Olympus, and Samsung. Kopin displays are used in camcorders, digital cameras, and the mobile video eyewear for watching TV, music videos, and movies; browsing the Web; and checking e-mail from mobile devices such as cell phones. If none of that seems a serious enough application, CyberDisplay LCDs are in the advanced night-vision goggles and thermal weapon sights used by the U.S. Army.

What Jacobsen is holding is only a prototype, and was released six months later as Golden-i™, a sleek, three-ounce, head-mounted display (HMD). "This device brings a person as close [to] the Borg collective as you can get without an operation," said Jacobsen, referring to the villainous cyborgs of *Star Trek*. This is by no means a "lite" technology. Golden-i features:

- Full-color high-resolution micro display
- Bluetooth, Wi-Fi, and cellular communication capability
- Six-axis solid-state head position and gesture tracker
- Texas Instruments' (TI) OMAP 3530 dual processor platform
- Microsoft Windows embedded CE 6.0 R2 operating system
- Optional image sensors: digital camera, including low light, IR, and the like
- Microphones, speaker, and ambient noise cancellation

"You could be standing between two race cars, gunning their engines, and have a normal conversation," said Jacobsen. "We've done a test like that."

In short, it is an always-on, always-connected hands-free computer. This is similar to technology now offered by MyVu, among others (see Figures 3.7 and 3.8). These interfaces are hands-free, and ultimately will be see-through, so that a user can see a display with one eye and keep walking or (saints preserve us) driving.

**FIGURE 3.7**
Your work station in 2025—or at least your telephone—will look a lot like this. (Courtesy MyVu)

**FIGURE 3.8**
The challenge of consumer-born high-fun interfaces is keeping them focused for the workplace. (Courtesy MyVu)

Michael Liebhold, author of *Infrastructure for the New Geography,* described a new physical-digital landscape that is emerging, which links places with "unprecedented amounts of information."[67] With wearable readers such as these head-mounted displays, an individual's physical reality becomes personalized, with cultural or historic information, or information about safety, weather, or public transportation, superimposed upon the landscape. This of course has enterprise applications as well; it is used in the military, in fighter jets and night-vision equipment.

Not all interfaces will be hands-free or pocket-sized. As Jacobsen observes, there is only so fast someone can type with two thumbs. Thus the comfortable terminal with QWERTY keyboard will remain, but likely, those terminals will be like ATMs or telephones; ubiquitous, and you can get service from any one of them. But you will require your own computing power, which will connect to these public stations in a Bluetooth-style Wireless connection. At its barest, your computer will be a sort of *Star Trek* "tricorder" that recognizes your voice commands. (Likely, your cellular phone already does that.)

And so, new forms of Wireless interfaces are inevitable. The standard issue military-grade always-on reality-enhancing interface you'd see on a fighter jet will make its way to civilians, and it will be made by the same companies. Swell, but what is it good for?

## Keeping the Interfaces on Mission

These interfaces are evolving to have full Web-browsing and cellular capabilities, meaning they are subject to all the forms of use and misuse as are smartphones and other computers. The challenge will be to keep them focused on mission-critical work.

Kopin foresees utility in logistics, distribution, military, and healthcare. For example, an HMD can enhance the performance and productivity of warehouse and distribution center staff; they need no longer check in at a dispatch station, or even pause to refer to a handheld mobile computer or dash-mounted lift-truck display. A truck driver or forklift operator can view high-resolution 15-inch images (i.e., high-resolution maps, facility navigation images, detailed documents, etc.) when traveling, and receive audio instructions with a full natural speech interface. Plus, the HMD travels with an operator, away from these vehicles, meaning the worker is "untethered."

The possibilities in healthcare are limitless. Nurses need no longer stop to record information such as temperature and blood pressure; they may simply say those numbers aloud. In time, that information will likely transfer to the HMD and the patient's file through a Bluetooth-style interface.

All of which is remarkable; but in the enterprise, these interfaces must be kept enterprise-grade.

Millennials are very clever in their use of Wireless Communication and Web 2.0, but they are poor self-observers and poor self-regulators. Enterprises must create rules surrounding the use of the technology, and use technology to enforce them, and the enforcement must be largely invisible and behind the scenes. But it is purposeless to try to cleave the work and social settings; we have moved toward this merged model for decades, and Web 2.0 has made it a reality (for good or for bad).

Ironically, like Lean and TPS, this model is somewhat Japanese; a Japanese company represented livelihood, friendships, status, and lifelong well-being to Japanese workers. But where the Japanese model was work and company centric, the Web 2.0 model seems to be more worker and socially centric. If workers can use HMDs to chat and access Facebook and YouTube, they'll do it. And if the HMD is capable of recording something idiotic said by a superior and shared, they'll do that too. Once again, in the Wireless/Web 2.0 world, users define what is proper use.

Thus, an enterprise must apply the same rules of recording and file sharing that we detailed earlier. Given that PMAs and HMDs are practical computers—futuristic-looking, powerful computers—they will require implementation. Here, just as with notebook computers and PDAs, the answer will be to focus the use upon the value stream, another mission for continuous improvement experts.

## Harness the Social Utility

If the Wireless/Web 2.0 world is socially centric, why fight it? As we observed earlier, Millennial learning is remarkably Lean. Millennials are quick to interact, quick to form communities, and quick to share ideas (as well as gossip and hilarious videos), using mobile computing to do it. If as we saw earlier this is a cost-effective method of training, then the social network is a perfect, readily usable, and low-cost mechanism for kaizen, A3, and knowledge bases.

*InformationWeek* editor Bob Evans observed that the social-media tool Twitter, "Gives businesses the unprecedented ability to tap into customer-driven feedback loops, which just on their own are highly valuable, and turn them into marketing labs, message amplifiers, focus groups, sales tests, and possibly even goodwill ambassadors."[68] Social networks also enable employees to share ideas and solutions, just as they do with kaizen and A3 forms.

As an example of serving customers, the Mayo Clinic used Twitter, Facebook, and YouTube to share a study on celiac disease (an immune-system response to gluten) with a few select patients. "Hopefully it becomes a model," said Mayo's Manager of Syndication and Social Media Lee Aase to *Forbes* magazine. *Forbes* observed that this model shatters the wall between patients and physicians; true, the patients do not get their answers directly from physicians, but they are satisfied with the just-released, first-round information that the clinic provides.[69]

If there is a company using social media within its four walls for A3 or kaizen, we have not heard of it. But we will.

## SCALABILITY: USE WIRELESS TO OPEN THE LOOPS

A fallibility of Lean is that it encourages practitioners to set the appropriate size for a process or staff based on current conditions, not future conditions. (Again, in essence, you buy the pants that fit you now, not the pants you wish would fit you.) In Wireless, scalability is a measure of success.

One key element of the Toyota Production System is that smooth production flow stretches beyond the four walls of a company (the "closed loop") to smooth the flow from suppliers and to the customers (the "open loop").

Closed loops are more finite in scope than an open loop. An old-style local area network before the Internet was a closed-loop system. Interdepartmental mail, in which someone typically wheeled around a cart delivering to "mail stops" was a closed loop system. An eKanban system on a modern shop floor, which connects the shop floor to the company's own stores, is a closed-loop system, and often good enough.

But, open the loop, and you open the Lean possibilities. With an open-loop eKanban system, your vendors see your replenishment needs coming, and deliver materials just in time; you maintain less inventory, because you use materials immediately. And if your open loop extends to your customers, you deliver goods immediately. Thus you have eliminated inventory on both ends of your process. This virtuous circle fosters collaboration and Lean thinking up and down the supply chain, ideally.

This requires solutions that scale up, and business is eager for those solutions. In its Annual RFID End User Survey 2008, ABI Research discovered that whatever the economic struggles of U.S. business, it remained bullish on implementing radio frequency identification technology because it can scale. More than 185 organizations from across the world responded to a survey of their RFID usage plans, adoption drivers, value propositions sought, and expectations of return on investment, or ROI.

Recall from Chapter 2 that among those companies, the four most important drivers for adoption of RFID were, on a scale of 1 being lowest to 5 being highest:

- Business process improvement (4.23)
- Ease of scalability/system extensibility (4.15)

- Ease of integration (4.07)
- Removal of human intervention (4.02)

The second and third drivers are ones of usability; these companies are anxious to use the technology and now find it cost effective and feasible. The first and fourth (process improvement and removing human intervention) are their top organizational priorities, and are perfectly aligned with the Lean and continuous improvement priorities.[70]

## Scalability, Enterprisewide

Recall that earlier in the chapter SAP's Ganesh Wadawadigi said, "Technology plays a key role in sustainability and scalability." Wireless implementations routinely begin as pilots with scalability in mind, targeting a systemwide problem. Recall as well that American Apparel achieved stellar results at its Columbia University location, and quickly scaled up to six more stores (and is in the process of implementing RFID systemwide).

Whatever its current health, Ford Motor Company in particular has led the way in achieving systemwide results with Wireless automation.[71] Ford first implemented its WhereNet Real-Time Location System in 1998, at its Sterling Heights, Michigan plant of 250,000 square feet. Building on that infrastructure, Ford Global Technologies (a subsidiary) and WhereNet co-developed WhereCall, a Wireless eKanban system, which is now at work in more than 35 Ford plants worldwide.[72] And it piloted WhereNet's Vehicle Inventory Management System (VIMS) in 2000, at its Michigan truck plant, with an output of thousands of vehicles daily. This proved so successful that it quickly rolled out VIMS at all its manufacturing plants in North America.[73]

Ford estimates that between more efficient use of labor and Wireless versus Wired technology, it has saved between $200,000 and $500,000 per facility; and reduced implementation time at each plant by several weeks over hardwired eKanban.

## Scalability into Supply Chain and Open Loops

Most of the superlative Wireless examples in *Thin Air*—from McCarran Airport in Las Vegas to Boeing to American Apparel retail stores—are closed-loop systems. The open-loop Wireless system for a distributed enterprise or for an extended supply chain is largely theoretical; sheer scope and complexity have kept it from happening.

The most immediate Lean Wireless opportunities are within the four walls: for an enterprise to improve its own processes, and minimize its own variable costs, before it attempts to improve its extended supply chain, or its vendor–supplier relationships.

This is because there are hundreds or even thousands of nodes of information on the extended supply chain. Let's take the example of a photocopier manufacturer, and list just a fraction of these, starting on the vendor side of the supply chain and ending at the customer:

- The salesperson's mobile computer tapping into the CRM system
- Both the Internet-based customer portal and vendor portal
- RFID on the raw material provider's outbound shipping dock
- GPS on the material provider's delivery vehicle (or third party's vehicle)
- RFID on the manufacturer's inbound shipping dock
- RTLS for key equipment at the manufacturer
- Machine-to-machine (M2M) sensors on the manufacturer's floor
- RFID on the work cells at the manufacturer
- PDAs and cellular phones on the shop floor
- RFID on the manufacturer's outbound shipping dock
- GPS on the manufacturer's or third-party's delivery vehicle
- PDAs for service personnel

Most enterprises are hard-pressed to integrate all the nodes within the four walls, much less from the extended supply chain. Even within the four-walls enterprise, cost, complexity, and a lack of standards keep the nodes from forming a chain. The ANSI/ISA-95 standard for manufacturers (the international standard for developing an automated interface between enterprise and control systems) uses the manufacturing execution system or MES as its backbone, but a supply-chain backbone must reach all of the nodes above and several hundred more.

Most WIP and visibility solutions offer no visibility beyond the four walls, which puts a cap on the benefits. A delay from a supplier requires immediate attention if production is to continue uninterrupted. The time of expected delivery is frankly too late to mitigate any losses; the balloon goes higher every minute.

Collaboration between suppliers and their customers is limited to purchase orders and advanced shipping notices; greater visibility into its customer's consumption enables a supplier to act before a PO is generated, with an "advance purchase notice"; in a retail setting, it would enable a manufacturer to send along a shipment well in advance of a stock-out; and in a clinical setting, to keep shelves stocked with critical medical supplies.

## The Requirement: Higher Visibility, Edge Intelligence

David Orain at Omnitrol Networks describes the requirements. "These new network-enabled ... devices need to be orchestrated at the edge of the network in order to support the ever-changing production processes without impacting the traditional ERP applications," writes Orain. "Finally, the solutions need to scale incrementally with an ultra-low total cost of ownership to accelerate adoption, large scale deployment and return on investment."

In addition, the systems must collect data from the edge, and deliver intelligence to the edge. The paradigm of unintelligent devices sending data to some server—in essence, a dumb infrastructure with a brain at the center of it—is giving way to smarter edge appliances with integrated application-ready middleware to make sense of it all.

## The Opportunity Finally Exists to Open the Loops

The opportunity to open the loops outside the four walls is not in the enterprise backbone; it is in the Web-based nerve path that connects nodes so well within the four walls. And, it is on the Internet, where people readily "hold hands," as Entigral's Bennet described it. Only a Web-based backbone can connect an enterprise's Wired and Wireless nodes, internal and external nodes, and everything and everyone, cost effectively.

## RTLS in the Open Loop

FedEx Ground (FXG) uses an Ekahau RTLS system for container tracking, and in a large and technically closed loop. But more than 500 locations is a demonstrably large loop.[74]

FXG uses Ekahau RTLS and Ekahau T301A tags to track and locate its high-value containerized mobile sorting docks across 30 ground hubs and more than 500 pickup and delivery terminals. FXG wanted to use its existing Wi-Fi footprint, and the ability to manage the containers from its Pittsburgh headquarters. This enabled them to avoid installing a new infrastructure at their hubs and terminals, and to track the containers by number of days at any location. With this intelligence, FXG is able to maximize the use of its existing containers, versus purchasing more.[75]

## CASE STUDY: KANBAN REPLENISHMENT[76]

A $30 billion durable goods manufacturer brought in Rush Tracking Systems to improve kanban accuracy and reduce hot calls (such as "where is this particular batch?") and rework, which it believed it could do by improving replenishment efficiency and accuracy.

What is intriguing about this instance is that it stretches the loop beyond the four walls of the plant, to include a third-party logistics provider and a third-party paint shop (Figure 3.9). This is not the largest of supply chains, but still is two steps beyond the closed loop or internal-only implementations we have examined thus far. This required RFID visibility at the original equipment manufacturer, the logistics provider, and the contractor.

The answer was to affix a durable RFID tag to the returnable totes of parts (see Figure 3.10). As the totes are reused, so are the tags. This makes for a compelling business case, as the cost-per-use of the tag is significantly

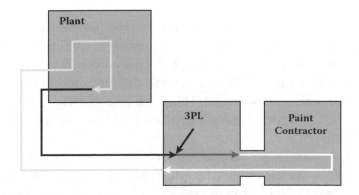

**FIGURE 3.9**
A miniature supply chain, but it needs to run flawlessly. (Courtesy Rush Tracking Systems)

**FIGURE 3.10**

An RFID tag on a returnable tote, tracing the tote (and parts) at the OEM, the logistics provider, and contractor. (Courtesy Rush Tracking Systems)

lower than the actual cost of the tag if it were used only one time. A single RFID portal at the third-party logistics provider (3PL) was sufficient to track the OEM-to-3PL transfer of totes, and a second RFID-enabled tunnel at the paint shop traced incoming totes.

The OEM used RFID-enabled lift trucks, with the Rush Tracking Systems' VisiblEdge™ solution integrated to the SAP enterprise backbone. It uses an industrial-grade RFID reader designed to withstand the shock and vibration that is common in industrial operations. The reader is attached to cargo antennas that are not only tough, but designed to read RFID tags in a nine-foot-high stack of pallets, basically what fits in the full height of a trailer. It quickly identifies a single pallet, double-stacked pallets, or even nine short pallets. Depending on the composite materials of the load, the contents can also be confirmed. For example, verifying paper and cloth materials being carried is no problem but a pallet of metal items can be tricky.

The durable goods manufacturer has its warehouse in Charlotte, North Carolina and its enterprise system (SAP) running in Chicago, Illinois. The two properties are connected using the secure corporate intranet. When an operator performs a move in Charlotte, the entire verification process takes no more than a few seconds before the operator has a confirmation from SAP on his vehicle-mounted terminal (VMT) screen that came back from Chicago.

The drivers no longer conduct any bar-code scanning. They receive directions on a vehicle-mounted terminal as to placement; the driver also receives a notification if the placement, or load, is wrong. This is a big change for operators as they can focus on driving and not scanning. It provides significant benefits and quality improvements in the process.

The system was automated to perform 20,000 scans per week along its supply chain, much as with the healthcare example we saw earlier. The performance improvements were:

- Hot calls were reduced from 52 per month to 0.
- The pilot started with two departments and 350 containers and has now scaled up to 10 departments and 6,500 containers.
- The company estimates a 28 percent ROI, predominantly from labor savings.

Clearly, Wireless has promise, and remarkable promise, but it also has peril: new, hitherto unknown forms of peril, and new forms of waste, even as it eliminates old forms of waste.

Yet the successes are countless. The difference between success and failure is *intent*, that the technology does not simply worm its way into an organization as a new and unavoidable tool. It is implemented with intent; a company knows exactly the problems it wishes to solve, and the results it expects, quantified in terms of hours saved, revenue generated, percentage of defects, and so on.

If all of that sounds familiar, it is because these are basic tenets of Lean and Six Sigma. Let's see what happens when organizations implement Wireless with intent.

## ENDNOTES

1. From a December 5, 2008 interview with Chris Warner of Motorola.
2. Hurwitz, S. 2009. Making things simple; The marketing of complexity. http://www.conference-board.org/articles/articlehtml.cfm?ID=374 (accessed January 9, 2009).
3. Hurwitz, S. 2009. Making things simple; The marketing of complexity. http://www.conference-board.org/articles/articlehtml.cfm?ID=374 (accessed January 9, 2009).
4. Hurwitz, S. 2009. Making things simple; The marketing of complexity. http://www.conference-board.org/articles/articlehtml.cfm?ID=374 (accessed January 9, 2009).
5. Websense, Inc. 2008. Enabling Enterprise 2.0. White paper.
6. Websense, Inc. 2008. Enabling Enterprise 2.0. White paper.
7. Websense, Inc. 2008. Enabling Enterprise 2.0. White paper.
8. Websense, Inc. 2008. Enabling Enterprise 2.0. White paper.
9. Websense, Inc. 2008. Enabling Enterprise 2.0. White paper.
10. CNN.com/technology. 2009. Thrift store MP3 player contains secret military files. http://www.cnn.com/2009/TECH/01/27/confidential.mp3.player/index.html (accessed January 9, 2009).

11. CNN.com/technology. 2009. Thrift store MP3 player contains secret military files. http://www.cnn.com/2009/TECH/01/27/confidential.mp3.player/index.html (accessed January 9, 2009).
12. Immediate ban of Internet social networking sites (SNS) on Marine Corps enterprise network (MCEN) NIPRNET. http://www.marines.mil/news/messages/Pages/MARADMIN0458-09.aspx (accessed September 9, 2009).
13. Websense, Inc. 2008. ThreatSeeker Technology Q3-Q4 2007 Results. White paper.
14. Kimball, R. and M. Ross. 2002. *The Data Warehouse Toolkit,* 2nd edition. Indianapolis: Wiley.
15. Heinrich, C. 2005. *RFID and Beyond.* Indianapolis: Wiley.
16. Heinrich, C. 2005. *RFID and Beyond.* Indianapolis: Wiley, p. 42.
17. Heinrich, C. 2005. *RFID and Beyond.* Indianapolis: Wiley.
18. Baudin, M. and A. Rao. 2005. RFID applications in manufacturing. http://www.mmt-inst.com/RFID%20applications%20in%20manufacturing%20_Draft%207_.pdf (accessed January 9, 2009).
19. http://www.rfidupdate.com/articles/index.php?id=1544
20. Liard, M. and S. Schatt. 2008. Annual RFID End User Survey. ABI Research, Inc.
21. Lo, J. et al. 2007. RFID in Healthcare: A framework of uses and opportunities. University of Arkansas Information Technology Research Institute (ITRI).
22. Womack, J. and D. Jones. 2005. *Lean Solutions: How Companies and Customers Can Create Value and Wealth Together.* New York: Free Press.
23. Heinrich, C. 2005. *RFID and Beyond.* Indianapolis: Wiley.
24. Heinrich, C. 2005. *RFID and Beyond.* Indianapolis: Wiley.
25. Wyatt, A. 2005. *Mobile Workforce for Dummies.* Indianapolis: Wiley.
26. Cornick, B. 2007. A 10-year roadmap for deploying RFID in medicine. *Pharmaceutical & Medical Packaging News,* http://www.devicelink.com/pmpn/archive/07/11/002.html, (accessed January 9, 2009).
27. Friedman, T.L. 2007. *The World Is Flat,* 2nd revised and expanded edition. New York: Farrar, Strauss & Giroux.
28. Rio, R. 2008. Wireless mobility enhances, Lean manufacturing and Six Sigma. ARC Advisory Group, Inc.
29. http://www.wired.com/wired/archive/12.07/shoppers.html?pg=2&topic=shoppers&topic_set=
30. ABI Research, Inc. 2008. Wi-Fi Real Time Location Systems (RTLS). http://www.digitalmediajournal.com/research/1001587-WI-FI_Real_Time_Location_System_(RTLS) (accessed January 9, 2009).
31. Smith, G. 1994. *Statistical Reasoning,* 3rd edition. New York: McGraw-Hill.
32. Background and summary of the California ePedigree Law. 2009, State of California. http://www.pharmacy.ca.gov/laws_regs/e_pedigree_laws_summary.pdf (accessed September 10, 2009).
33. Counterfeit drugs kill! 2007, World Health Organization. White paper.
34. Taylor, N. 2008. Another delay knocks back ePedigree to 2015. *in-Pharma,* October 13.
35. Background and summary of the California ePedigree Law. 2009, State of California. http://www.pharmacy.ca.gov/laws_regs/e_pedigree_laws_summary.pdf (accessed September 10, 2009).
36. Womack, J. and D. Jones. 2005. *Lean Solutions: How Companies and Customers Can Create Value and Wealth Together.* New York: Free Press.

37. Womack, J. and D. Jones. 2005. *Lean Solutions: How Companies and Customers Can Create Value and Wealth Together.* New York: Free Press.
38. Rutanen, T. 2008. *Asset and People Tracking with Wi-Fi RTLS.* Ekahau, Inc. White paper.
39. Rutanen, T. 2008. *Asset and People Tracking with Wi-Fi RTLS.* Ekahau, Inc. White paper.
40. Rutanen, T. 2008. *Asset and People Tracking with Wi-Fi RTLS.* Ekahau, Inc. White paper.
41. Interview, 12 February 2008, Ralph Rio, ARC Advisory Service.
42. Rutanen, T. 2008. *Asset and People Tracking with Wi-Fi RTLS.* Ekahau, Inc. White paper.
43. Aberdeen Group. 2007. Can RFID deliver the goods? The manufacturer's visibility into supply and demand. White paper.
44. http://newsroom.accenture.com/article_display.cfm?article_id=4767
45. http://newsroom.accenture.com/article_display.cfm?article_id=4767
46. Inayatullah, S. 2006. Eliminating future shock—The changing world of work and the organization. Futurist.com.
47. Inayatullah, S. 2006. Eliminating future shock—The changing world of work and the organization. Futurist.com.
48. Inayatullah, S. 2006. Eliminating future shock—The changing world of work and the organization. Futurist.com.
49. Jenkins, H. 2006. *Convergence Culture: Where Old and New Media Collide.* New York: New York University Press.
50. http://www.businessweek.com/innovate/content/feb2007/id20070213_645807.htm
51. http://www.businessweek.com/innovate/content/feb2007/id20070213_645807.htm
52. Casner-Lotto, J., E. Rosenblum, and M. Wright. 2009. The ill-prepared U.S. workforce. The Conference Board.
53. Casner-Lotto, J., E. Rosenblum, and M. Wright. 2009. The ill-prepared U.S. workforce. The Conference Board.
54. Casner-Lotto, J., E. Rosenblum, and M. Wright. 2009. The ill-prepared U.S. workforce. The Conference Board.
55. Casner-Lotto, J., E. Rosenblum, and M. Wright. 2009. The ill-prepared U.S. workforce. The Conference Board.
56. http://www.futurist.com/articles/future-trends/eliminating-future-shock/
57. Lang, S.B. 2004. The impact of video systems on architecture, dissertation, Swiss Federal Institute of Technology.
58. Bentham, J. 1995. *Panopticon (Preface).* In Miran Bozovic (ed.), *The Panopticon Writings,* London: Verso, pp. 29–95.
59. Cascio, J. from a December 4, 2008 interview.
60. Perez-Pena, R. 2008. A senior fellow at the institute of nonexistence, *The New York Times,* November 12, http://www.nytimes.com/2008/11/13/arts/television/13hoax.html (accessed August 27, 2009).
61. http://en.wikipedia.org/wiki/Encyclopedia_Britannica#cite_note-86
62. http://newsroom.accenture.com/article_display.cfm?article_id=4767
63. http://www.worldchanging.com/archives//002651.html
64. http://www.worldchanging.com/archives//002651.html
65. http://robotic.media.mit.edu/projects/robots/huggable/overview/overview.html

66. http://www.worldchanging.com/archives/002651.html
67. http://www.iftf.org/system/files/deliverables/SR-869_Infrastructure_New_Geog. pdf
68. Evans, B. 2009. Global CIO: Why CIOs need the transformative power of Twitter. *Information Week*, August 3.
69. Evans, B. 2009. Global CIO: Why CIOs need the transformative power of Twitter. *Information Week*, August 3.
70. Liard, M. and S. Schatt. 2008. Annual RFID End User Survey. ABI Research, Inc.
71. ZIH, Inc. 2008. Ford Motor Company. http://zes.zebra.com/customers/cs-ford.jsp (accessed January 9, 2009). Case study.
72. ZIH, Inc. 2008. Ford Motor Company. http://zes.zebra.com/customers/cs-ford.jsp (accessed January 9, 2009). Case study.
73. ZIH, Inc. 2008. Ford Motor Company. http://zes.zebra.com/customers/cs-ford.jsp (accessed January 9, 2009). Case study.
74. Rutanen, T. 2008. *Asset and People Tracking with Wi-Fi RTLS*. Ekahau, Inc. White paper.
75. Rutanen, T. 2008. *Asset and People Tracking with Wi-Fi RTLS*. Ekahau, Inc. White paper.
76. Rush, T. 2008. Using RFID to drive operational efficiencies. Webinar presentation through the Society of Manufacturing Engineers.

# 4

## *The Democratic Frontiers of Lean Wireless*

The Wireless camp has never been in awe of Toyota, or the Toyota Production System, which are no longer the only measures of a well-run enterprise. We now have cost-cutting and supply-chain champions such as Hewlett-Packard, Procter & Gamble, and Hyundai, which as of this writing, is weathering the economic downturn far better than is Toyota.

One Wireless pioneer to whom we spoke (who would only allow us to quote him anonymously) sees the Lean/Toyota connection as downright quaint. He described Lean zealots as "A bunch of old farts trying to find a better way to make a Toyota. I love those guys, but Lean got started in automotives, a big industry with relatively low volume. Apply it to the higher-velocity supply chain, and you'll see remarkable things happen."

Best practices in Lean Wireless can come from any vertical, and any enterprise, large or small; and they are more easily identifiable, scalable, and transferrable than enterprise advances of old. Lean and Six Sigma both trace their roots to manufacturing—Lean to Toyota, Six Sigma to General Electric—and it's in manufacturing that we've seen its greatest successes. But as we show in the coming pages, Wireless best practices are more democratic. In this chapter, we explore best practices from the Thames Valley Police, the Obama campaign, from libraries, and the cath lab at Mercy Medical, as well as from manufacturers such as the old Daimler Chrysler and logistics providers such as FedEx.

Moreover, every industry, company, and individual continuously improves Lean Wireless. Lean is a practice, with improvements made by end users where work is done. Because Wireless is a technology, its improvements are driven by its users and technology providers. The users are both consumer and commercial, which is encouraging; both amateur and commercial users, for example, contribute enthusiastically to open-source solutions such as Linux. Finally, every business and every consumer uses Web 2.0, which has created a seamless conduit for communication and innovation.

## CROWDSOURCING; WE ALL OWN (AND DRIVE) INNOVATION

We are at a unique period in history. A great idea can come from absolutely anywhere, and it can be communicated instantly, something we take for granted. For centuries, and up until air travel and telephony, separate geographies owned specific schools of innovation. During the Renaissance, advances in architecture came from Italy. In the Victorian era, advances in surgery came from the Royal College of Surgeons in Edinburgh, Scotland. Geographies specialized, and people who wished to innovate in architecture or surgery went, as students, where the innovation was welcomed. Then the grips loosened, and companies owned innovations. In the 1950s, any advance in business computing was likely made by IBM, but in time, Digital Equipment, Hewlett-Packard, and Xerox, among others, created a competitive market.

Now, no country, company, or city owns a particular school of innovation. Innovation is plentiful and unceasing, and cooperation is not bound by geography. Two individuals can collaborate virtually on a product design, one in Silicon Valley and one in Mumbai. Similarly, two people can spend hours gathering a team to smash a bullying troll in the online game *World of Warcraft*. Granted, one application is more constructive than the other, but they work the same way.

Both bad and good ideas come from crowds. Crowds are evil or indifferent at their worst, heroic and insightful at their best. Lynchings in the early twentieth century were groupthink turned violent; these frequently began with a spontaneous minor incident (such as the black boy who whistled at a white woman) or a rumor that caught fire and turned into a spontaneous bloody party.

The more constructive cousins of groupthink and the bystander effect are "crowdsourcing" and "open innovation." George S. Day, co-director of the Wharton School's Mack Center for Technological Innovation, dubbed open innovation "crowdsourcing," which entails collaborating with partners to solve business problems.[1] Mack points to the Waltham, Massachusetts-based company InnoCentive, a sort of eBay for companies with science, engineering, or business challenges. In essence, problem solvers worldwide "bid" to solve the problem. In one real-life example, a company (or "Seeker") is, to quote its request for proposals, "looking for a material that changes color upon submersion in water. Current commercially available technologies are typically inadequate in some environments." The company is offering a $30,000 "Challenge Reward." Elsewhere, a Seeker is offering $40,000 for a technology to preserve the texture of freshly baked bread upon extended storage. Forty grand is not, frankly, top dollar for such an idea.

General Mills, which introduces more than 300 new products per year worldwide, turned to crowdsourcing when it created its Worldwide Innovation Network, or WIN. The food industry is historically slow to innovate. In 2004, Senior Vice President Peter Erickson observed how dynamically biotech and pharma companies reached out to collaborators, and tasked his own company to do the same. General Mills announced that it sought innovation, and invited prospective partners to approach them directly through a specialized department, or through a dedicated Web portal. (Erickson himself entertained prospective partners at speaking engagements.) A partner could be a giant such as DuPont, or a lone citizen with a great idea.

One conspicuous example was its Nature Valley Fruit Bar: General Mills required healthy fruit snacks for sale in club stores like Costco, and a third party approached the company with its formulation, to which General Mills made only minor adjustments. General Mills called not only for product, but for innovations as well, and partnered with MTL Technologies in Montreal for a portable thermoelectric cooler design; the design allowed General Mills to place Pillsbury pie crusts, for example, in the baking aisle of stores. As Public Relations Manager Kirstie Foster told the authors in a 2006 interview, the advantage to WIN applicants is in joining with strong established brands such as Betty Crocker or Yoplait, and General Mills in turn offers "the scale, credibility, and scope to help our partners' business plans take off."

The Americas' SAP User Group, or ASUG, is a strong model of crowdsourcing; the organization has since the early 1990s partnered with enterprise software giant SAP as advisers, making their requirements known, and sharing best practices among themselves. And Linux, described as an "open-source code," is really crowdsourced, and the crowd is overall very pleased with the results.

Lean Wireless is crowdsourced and Web 2.0 is its platform. Let's examine how success transfers across verticals, across geographies, and throughout hierarchies.

## THE WIRELESS LIBRARY

According to *RFID Journal*, more than 2,000 libraries worldwide have installed RFID systems, chiefly for:

- Self-service check-in and check-out
- Protection against loss by theft
- Inventory accuracy and control
- Automated sorting and cataloguing

Singapore, Germany, and The Netherlands have all chosen RFID as a technology of choice to efficiently manage new libraries or libraries that receive new IT infrastructure. In Singapore, for example, libraries use more than 15 million RFID tags from NXP Semiconductor.[2]

If libraries seem a low priority for Lean consultants, consider the characteristics of a library system:[3] it is a closed-loop system (as are most enterprises), it maintains a small amount of inventory (including books, CDs, DVDs, and journals), and Internet connectivity allows interlibrary transfer very efficiently, thus opening the closed loop to multisite capabilities. It is a short leap between a statewide library system, and a multisite enterprise such as Walmart.[4]

## THE WIRELESS ATHLETE

Beginning in April 2007, runners in the highfalutin Boston Marathon were required to attach RFID tags to the laces of their running shoes.

With more than 23,000 runners competing in 2007, Boston had yet to figure out how to eliminate cheating; Boston was famously rocked by scandal in 1980, when a fleshy unknown named Rosie Ruiz sailed across the finish line ahead of better-known competitors such as Canadian Jackie Gareau.

Ruiz had hopped a subway train and skipped several miles of the race. She was busted not by any technology, but by eyewitnesses. For the next 26 years, video surveillance provided some interim solution, but it wasn't until 2007 that the Boston Athletic Association (BAA) could conclusively verify that each runner passed every checkpoint.

Finally, RFID allows for a richer experience for supporters. The tags transpond at five-kilometer intervals on the BAA Web site, and send progress alerts to e-mail addresses and Wireless devices. Wireless and automatic identification similarly enables FedEx customers to track packages. This same utility is at work in security and in time and attendance, verifying that security guards make rounds and employees report for work without punching clocks; employees and security personnel need not actively punch a clock, rather, their presence is recorded by RFID tags (usually encased in a badge) and readers.

## EXTREME TRACEABILITY, FROM FARM TO FORK

One of the most widely known RFID initiatives for food traceability projects is with the State of Hawaii Department of Agriculture and the Hawaii Farm Bureau. The Hawaii Produce Traceability project has achieved "farm to fork" traceability by using passive RFID tags to track and trace fresh produce throughout the state's food supply chain. The first system of its kind in the United States, it's designed to promote food safety by providing product visibility down to the farm or even field level. The RFID system provides detailed, real-time information, which is used to improve inventory control, optimize the supply chain, and most important, enable recalls in less than an hour.

In the first phase, Lowry Computer Products developed a solution using RFID readers from Motorola and waterproof, passive, RFID smart labels to provide real-time supply-chain data that shows when boxed produce is planted and harvested, what pesticides are used, and when and where

RFID-tagged boxes are scanned. The data is automatically uploaded into a database, where it can be used by such program participants as distributors, retailers, and restaurants. It is also available for public review on the initiative Web portal, hawaiifoodsafetycenter.org.

Growers place the waterproof RFID labels on boxes of produce and record the information using an RFID handheld system. The same boxes are read upon entry to the distribution center and again on exiting both the physical facility and cold storage. Tags are read again at the retailer's point of entry, removal from cold storage, and at end of life, when expired product is dumped. Both the distribution center and retailer use stationary RFID readers in portals.

Participants can use the data to optimize harvest productivity, strengthen food processing controls, increase cold chain visibility, reduce produce dwell time on shipping and receiving docks, accelerate transportation times, and improve inventory turns. This enables them to optimize margins in the competitive food industry. In the event of a food recall growers can quickly identify if they are affected, thus enhancing their brand and protecting revenues. Affected growers can localize the impact of relevant recalls to the field level, minimizing losses.

State officials are considering enhancements to the next two phases of the project, such as deploying RFID-enabled cell phones to enable more farms to participate, and implementing produce temperature tracking to reduce the threat of food spoilage. The initiative could be expanded to cover 5,000 state farms at full implementation.

> "The Hawaii Produce Traceability initiative is an integral part of the State Food Safety Certification system," said Dr. John Ryan, Administrator, Quality Assurance Division, State of Hawaii Department of Agriculture. This project provides the backbone for future and more preventive closed-loop sensor technologies which are capable of measuring and reporting biocontaminants and temperature variations via the RFID system as produce moves through the supply chain. The RFID system will provide managers with improved real-time control over potential food safety problems and help to prevent wide-spread human and economic impact.

When it comes to recalling or controlling accidentally or intentionally contaminated food, the capabilities of this real-time technology provide never-before-possible protection to consumers.

## THE LEAN WIRELESS ELECTION: A SUPPLY CHAIN SUCCESS

> If you don't know who is handling election machinery and data, there is absolutely no way you can know whether your election has been either fair or accurate.
>
> **BlackBoxVoting, "America's Elections Watchdog Group"**

Cutting delivery times from five hours to two, with far less physical labor and near 100 percent accuracy, would be a success in any supply chain. The Registrar of Voters in Alameda County, California achieved those results with RFID, and led the nation in securing the results and improving the logistics of polling.

The cranky 2000 presidential election (George W. Bush versus Al Gore, which was decided in Florida) left Americans dismayed; paper balloting seemed primitive and insecure. So did the logistics of voting, which called for dragging equipment and ballots to and from polling places. Rumors abounded of trash bags of ballots found by Florida roadsides (presumably votes for Gore, although no such trash bags were ever produced). CNN observed that without a secure chain of custody, someone could swap memory cards with ballot results before tallying, and that drop-off points lacked security and organization. Electronic polling seemed the answer, but a grand-scale, perfectly secure system was not yet practical.

Still, electronic methods would ensure that all votes were counted, and counted in a timely way. Dave MacDonald, director of information technology and the registrar of voters in Alameda County, California, contracted with RFID Global Solution (RFIDGS) to devise a first-of-its-kind system called SecureVote™. RFIDGS develops wireless asset and personnel management systems for industry clients and government agencies including the Department of Defense and NASA.

The paper balloting process did involve built-in security, including involvement of sheriff's deputies and tamper-proof locks and seals on polling equipment, with no security breaches recorded. Security was less the challenge than logistical efficiency. The voting process in Alameda County involves about 75 polling places for a local election, and more than 850 for a presidential election.

Each polling place received a serialized canvas bag with tamper-proof seal, which contained three components: a memory pack, holding voting results scanned from paper ballots; a personal computer memory (PCMCIA) card (much like a Flash Card or Memory Stick), holding voting results from touch screens used by disabled voters; and a paper voting roster. Each component was tagged with a bar code joining it to a "parent" bag. The bags went to about 30 collection points which serviced the polling places (typically community centers, churches, and high schools).

When polls closed, the same exercise happened in reverse; sheriff's deputies would pick up and deliver the bags to the Vote Count facility in Oakland, California, and then to the central distribution warehouse. Dozens of volunteers unzipped each bag to determine if the three components were present; this could take four to five hours. All the bags arrived nearly at the same time, and the bags stacked up quickly. If a component was missing, "We may not know until after midnight and then we have to find it," said MacDonald, which is a miserable process that can go on until 4 a.m. A missing component required sending someone to whatever church basement or high-school gymnasium served as a polling place and waking up a precinct captain and a custodian or someone else who could open the polling place.

In mapping the current voter server system, RFIDGS identified three key read points along the chain of custody:

- The central distribution warehouse, which RFIDGS likened to a store-level distribution center
- Intermediary collection points
- The Vote Count facility

Now, using SecureVote, each component is tagged with an ISO-18000-6C passive (non-battery-powered) RFID tag at the central warehouse, and is joined to a canvas bag with parent RFID tag. County employees use Motorola MC9090-G RFID handheld readers to verify outbound shipments, and precinct captains pick up the bags at the warehouse.

After polling, the bags are brought to collection points (typically police stations) where a Registrar of Voters employee uses another Motorola handheld reader to scan each bag. Previously, the tamper-proof seal had to be broken, the contents verified, and then resealed. Now, a bag that registers no errors need not be opened at all, which greatly improves security.

Next, sheriff's deputies bring the collected bags to the Central Tally facility, where more deputies pass the bags through a reader portal, which includes a Motorola stationary reader and four antennas. The portal verifies that all contents are present, and a volunteer simply monitors the results on a screen, giving a "go" signal or directing the deputy to a help desk if a component appears to be missing (signaled by an "X" over a visual of the bag).

The help desk double-checks the bag contents in two ways: with an RFID reading, and by physically opening the bag. The help desk uses the RFIDGS SmartTable™, a plug-and-play RFID reading platform with a 6-foot by 3-foot workspace. SmartTable includes a reader, antennas, network interface, and light integrators. It is also used by the U.S. Department of Defense to manage parts and ammunition and at Homeland Security checkpoints. Most often, the missing component is present; there was a misread at the first checkpoint, or a tag may have been damaged.

In the first test, which was a local election involving 75 bags, tag reads were 97 percent accurate, and missing components from 10 duffel bags were located within six minutes, a tremendous improvement over the four hours necessary before the implementation of the SecureVote system.

The November 6, 2007 polling was near perfect. Polling closed at 8 p.m. and employees were able to read and verify the contents of every bag in an hour and 20 minutes, compared to the four or five hours that was typical using bar codes. The bags were dispatched to the warehouse, and the votes were tallied before 11 p.m..

"I had high expectations and this exceeded them," MacDonald explained after November 6, 2007. "In 2008, California has three major elections, which hasn't happened since 1948, and the security requirements are tougher than they have ever been in history." Alameda County contracted RFIDGS to scale the system by 1,000 percent for the presidential primary, which involved 825 precincts and 27 return centers. MacDonald summarized by saying, "We have a low-tech front-end with paper and pencil voting, but a high-tech back-end with real-time tracking and security to account for everything. My goal is to ensure everyone's vote is counted."

## THE LEAN WIRELESS ELECTION 2:
## WEB 2.0 AND DECISION 2008

In the 2008 presidential election, young volunteers stood in polling places, listening carefully as voters declared themselves. The volunteers phoned in the names of the voters to Obama and McCain headquarters, or communicated them with e-mails from smartphones, and those names were struck from databases by other volunteers, so that further volunteers would know who had yet to vote and so whom to phone ("May Mr. McCain/ Mr. Obama count on your vote today?"). Of course, it would have been even more efficient had the registrars of voters simply made a live database available in real-time to the Obama and McCain camps, but registrars of voters weren't about to do that.

The Obama campaign had a very clear and very strong Wireless and Web 2.0 strategy from its inception. Like an Andon board, the Obama camp disseminated positive news and polling statistics daily, to wherever a supporter might listen. Among other measures, they:

- Posted campaign messages on YouTube
- Used supporters' cell phone numbers for regular text message announcements (including Obama's selection of Vice President Joseph Biden)
- Sent regular updates and messages via Twitter, Facebook, and MySpace, among other social networking tools[5]

Every one of these applications becomes a Wireless one, if used on a BlackBerry, iPhone, or other smartphone. Twitter in particular is a wireless application. It is used largely on cellular phones and PDAs, for instant messages on an account owner's comings and goings. They are usually frivolous ("JennyB is at the dentist with an abcess OMG :(") and invite users to keep track of friends and Internet-only celebrities such as self-promoter Lisa Nova. The Obama camp used Twitter to keep followers informed of the candidate's movements, meetings, and endorsements, and the momentum seemed endless. He is also credited with the single highest number of Twitter "followers" of any account.[6]

Supposedly, Obama led with 844,927 MySpace "friends," versus 219,404 for McCain at the end of August 2008 (although one of the co-authors was listed as one of those friends, despite never visiting the Obama account).[7] Numbers such as those—Obama's superior MySpace rankings versus McCain and Senator Hillary Clinton—made the news on CNN and NPR, and in *The New York Times* and *WIRED* magazine.[8]

If the leap in utility from electioneering to Lean and continuous improvement seems large, consider the statistic that messages created by the Obama campaign generated a measurable 14.5 million hours of viewing, equal to an estimated $46 million in network time.[9] It cost the Obama campaign nothing to upload those videos. That is as good an example of doing more with less as was ever achieved in industry, but it is a form of marketing, which is off the "value stream."

Marketing might be considered one of the structural costs, the "costs of doing business" that ultimately increase the cost of goods sold; but marketing cannot achieve too much, and cost too little. All of the major auto suppliers have Facebook pages, inviting enthusiasts and potential buyers to trade information on new models. BMW pioneered Internet marketing with short films directed by hotshots such as U.K. director Guy Ritchie on its Web site; now, keying in BMW on YouTube reveals more than 100,000 BMW-positive films. Searching Hewlett-Packard on YouTube reveals an interview with CEO Mark Hurd and instructional videos for HP consumer products (chiefly printers), that are on the value stream, satisfying the customer with do-it-yourself support, where the customer is used to finding information. The short story is that Web 2.0 is far from trivial. Used right, it is Lean in cost and highly effective in messaging.

## LEAN WIRELESS PRESIDENT AND POLICE

President Obama famously struggled with surrendering his BlackBerry. "His BlackBerry was constantly crackling with e-mails," on the campaign trail, said David Axelrod, the campaign's chief strategist.[10] Obama believed strongly that Wireless technology is a modern tool of management, and it seemed absurd to him that the President of the United States would be disallowed the most modern tools. Still, on top of any threats about e-mail security, Obama is subject to the Presidential Records Act, meaning any correspondence is subject to the Freedom of Information Act, and that's a practical point: within a week of Obama's inauguration,

U.S. District Judge Henry Kennedy instructed officials to search all White House workstations in the Bush White House, and to "collect and preserve all e-mails sent or received between March 2003 and October 2005." This was in answer to an ongoing lawsuit by public interest groups alleging failure by the Bush White House to monitor internal communications among staff. The Bush White House is expected to transfer more than 300 million e-mail messages and 25,000 boxes of documents (exclusive of any missing) to the National Archive.[11]

BlackBerry got more positive press from Obama's endorsement than it did as an enterprise tool at JP Morgan Chase, Home Depot, or Coca-Cola. It generated more excitement when Paris Hilton supposedly hacked Lindsey Lohan's BlackBerry to send out embarrassing and insulting e-mails under Lohan's name, than it did as an A3 form on the shop floor at Boeing.

These applications do not simply make the leap into industry. Research In Motion (RIM), maker of the BlackBerry PDA, understands this very well. RIM offers the Blackberry Enterprise Server to integrate the device into a company's backbone, and counts among its partners SAP America, Sybase, and PeopleSoft, all of which offer interfaces to the BlackBerry Enterprise Server. RIM famously partnered with salesforce.com in 2003, to deliver on-demand customer relationship management (CRM) applications to BlackBerry devices. Using the BlackBerry, JP Morgan Chase employees are able to work with customers on site, versus requiring customers to come to a branch to complete, for example, a loan application.

In another vertical, the Thames Valley Police in the United Kingdom and Cape Breton Regional Police Force in Nova Scotia, Canada use BlackBerry smartphones for:

- Accessing missing persons databases
- Accessing warrants and court order databases
- Neighborhood crime notifications
- Witness and custody video and photographs

"A big plus is we're starting to e-mail mug shots to our officers who might be on a stakeout or on patrol," said Inspector Thomas Hastie of the Cape Breton Regional Police Service. "They can compare the image on their BlackBerry smartphones to suspicious individuals, which means they're safer when they approach someone."[12] The Service uses the OnPatrol application from XWave, written specifically for police forces. Using this application, the Thames Valley force significantly reduced the 30 percent of an officer's day spent at the station completing administrative tasks.

Police officers and JP Morgan salespeople are on "the edge," as much as are the production workers at Boeing, or the service personnel at Whirlpool. In its early days, BlackBerry was a status symbol, and largely an executive tool; but as the platform evolved, it has made its way to those workers on the edge.

Where BlackBerry is most successful is where it is implemented with intent; JP Morgan knew exactly what improvements it required using the BlackBerry, as did the Thames Valley Police Force, and it can measure those improvements. Otherwise, a company risks dividing its employees' attention with the ceaseless connectivity of the "CrackBerry" (as the devices are nicknamed).

That risk aside, BlackBerry earns its place as an enterprise device. As of the second quarter of 2008, RIM captured more than 50 percent of the smartphone market in the United States, a leap from 44.5 percent in the first quarter, according to IDC Research.

RIM has always had an eye on both the business and consumer markets, and offers numerous models; the Apple iPhone, by contrast, lost sales in the second quarter as customers waited for the next-gen iPhone. Apple apparently has no more of a cohesive corporate-customer strategy for the iPhone than it had for the Macintosh computer. This is one instance, in Wireless adoption, wherein the consumer effect has little muscle; however popular the Mac is among college students, its statistical adoption in commercial applications is limited to smaller enterprises and art departments.

## WIRELESS JUSTICE

If you've watched *Law & Order*, you've seen some poor detective with a big sub sandwich and a large coffee, settling down to watch hours of videotape from a convenience store or jewelry store holdup, trying to spot a clue in a single frame of footage. Agonizing.

Wireless enables a detective or security officer to see when and where an item was last seen—even if it is hidden in a briefcase or under a jacket—and identify who had it, by matching movement data about the item to video footage.

Of course, not all missing goods are stolen goods. As we touched on in Chapters 1 and 3, the State of Florida implemented SimplyRFID's NOX system on file folders in its court system. NOX combines passive RFID with overt and covert high-def video surveillance. Again, one Florida county alone has 40,000 criminal cases per year, and relies upon physical files (most legal systems do, and they have to, as copying those documents is usually disallowed). If a file folder goes missing when a case comes to the docket, a judge might hold off for 15 minutes while such high-value talent such as district attorneys and paralegals scramble to find it. And if they don't, a defendant walks.

Now, an IT professional can pinpoint the last known location of the file in question. In one real case, that file was found in the mailroom and in transit, and in another, at the bottom of a paralegal's stack of files. A second use is in sting operations and surveillance (see Figure 4.1), and here is where the FBI uses NOX.

In 2006, the FBI was investigating thefts in a warehouse which trafficked high-value electronic goods such as plasma TVs. This warehouse processed more than 15,000 shipments, and video alone was impractical to spot an item that might be hidden inside another carton, for example.

**FIGURE 4.1**
RFID plus video surveillance. (Courtesy Simply RFID)

The FBI placed $.20 tags on all the high-value items in the warehouse, and combined with video surveillance, had the evidence to ID the thieves.

Finally, Wireless can secure such corporate assets as PCs and laptops. In an organization with hundreds, even thousands of computers, it's important to monitor what goes out the door, but impractical to sound the alarm every time a laptop passes a checkpoint in the normal course of business. If an item is missing, asset surveillance allows a security guard to call up a time and location when it was last seen.

As SimplyRFID CEO Carl Brown describes, laptops are good for about $100 at a pawn shop, but are not the only corporate assets worth taking.

Anything that someone can resell or use at home is likely to disappear. Think of $100 printer cartridges, or $5 reams of paper. A stapler might seem a trivial item to tag, but think of it as a "gateway" theft. Someone who'd steal a stapler is likely to steal something more valuable, given the chance. As Brown told the authors,

> Now think beyond corporate assets, to artwork, auto parts—once again, anything not nailed down and with some resale or personal value. Right now, copper wire theft is running at millions of dollars per day, because of the resale value of copper.

Of course, not all these cases involve sinister motives. In the case of the State of Florida, neither the mailroom clerk nor paralegal had any criminal intent in mind, nor do most people who leave their offices with laptops in their bags. At a distribution center, someone who was supposed to load four plasma TVs on a truck might load five by mistake, but you can trace the one that's missing to that dock door and see who was driving the lift truck.

One very clever use of asset surveillance is in tool tracking. Recall from Chapter 3 that Ford, DeWALT (of toolbox fame), and RFID provider ThingMagic have partnered to offer RFID-enabled pickup trucks, to verify that all expensive tools have been loaded back into the truck at the end of a job, or that the right tools are loaded before a job.

## THE SIGNIFICANCE OF iWIP

Lean emphasizes continuous productive flow, which is a requirement in every enterprise, be it a hospital, theme park, or retail store.[13] Recall that in

Chapter 3, we concluded that Lean's new focus must be on moving things, versus making things. Work-in-process, or WIP, is a form of movement, and ideally, WIP is tracked electronically, without interrupting flow. At its best, Web-based work-in-process visibility (iWIP) provides:[14]

- Work time and wait time at each stage, both actual and average
- Estimated completion time
- Location of a given work order or status

WIP is an Operations matter, not an IT matter. With WIP, you can access feedback on operational efficiency, including:

- Estimated completion times
- Tracking rework and costs
- Average timing by worker and workstation

The Rautakesko Oy hardware chain, the largest in Scandinavia and the Baltics, uses foot traffic analytics to maximize product placements and customer experience.[15] Rautakesko uses Ekahau RTLS tags on its shopping carts to monitor customer traffic patterns in its store, recording where they stop, where they aggregate, and which typical paths they take.[16] The display is configured like an infrared heat map, with red representing the highest traffic over a period of time; green representing medium traffic; and blue being cold, with little traffic.

Sometimes, the requirements of iWIP are much simpler. CVRD Inco wanted to track productivity at its Stobie Mine in Ontario, where the company mines nickel. Productivity was easily measurable, by tracking vehicle trips by loaders to crushers; each load measured approximately two tons of ore. The driver would record the trips on pencil and paper, calculate tonnage based on the number of trips, file the paperwork, and after perhaps two days, management would have an approximate tonnage. This of course was a chore to drivers, and subject to the usual errors of paperwork; besides that, it slowed production.[17]

CVRD Inco wanted more immediate and precise measures. It wanted to track and locate vehicles on a common communications infrastructure, and to make use of its existing infrastructure, which was an 802.11 a/b/g minewide Wi-Fi system. (This is in essence an "industrial-strength" version of the same Wi-Fi technology used in coffee shops and home networks.)

The company implemented a system of hardened Ekahau T-201 rechargeable tags mounted on its mining vehicles. The system provided a real-time map-based view of the mine and equipment location, and real-time measures on a Web-based dashboard. The turn-key system was implemented mostly by Inco employees, rather than by Ekahau or by third-party integrators.

The system was configured to measure and monitor production of ore by tracking vehicle trips to the crushers. This allowed the company to measure productivity in real-time and catch bottlenecks and process faults when they occurred. It also enabled them to reward employees for superior service. "Any time the front loader comes to the crusher, it has a full load," describes Ekahau's Tuomo Rutanen.

> The miners are incented on daily volume production. If they work an eight-hour shift, and they get an additional bonus based on productivity. They won't be down there playing poker; [iWIP] lets them work hard and crank stuff out of the ground, and they're compensated for it and automatically measured. Getting the scoop runs going and measured and accounted correctly makes sense for all parties, Rutanen explained.

## iWIP IN LEAN MANUFACTURING

"A tagline we use is, 'We make the shop floor talk'" Raj Saksena of Omnitrol Networks told the authors. "We believe that the information you get, and how you handle it, should not come from *people* working on the shop floor, rather, it should come *from* the shop floor."

That was the requirement at Endwave Defense Systems (now part of Microsemi RFIS), a company that actively practices Lean manufacturing, and uses Wireless technology to achieve it. Its key customers, including the Department of Defense, required more than quality and on-time delivery of equipment; they required visibility into the Endwave shop floor, to track order status.

Its 22,000-square-foot agile manufacturing plant near Sacramento, California produces radio frequency modules (including synthesizers and oscillators) for homeland security, defense electronics, and telecommunications networks. Endwave uses the Lean practices of cell-based production flow and kanban replenishment to keep orders flowing steadily.

Still, Endwave had limited visibility into WIP, work orders, and production bottlenecks; instead it relied upon gemba walks by production managers.

The solution was iWIP, from a talking shop floor. Visibility is a strong market driver for Wireless solutions, but as Saksena observes, "Visibility only gives you so much value. If you apply intelligent services around visibility, you get a whole new avenue of optimizing performance." Tracking by WIP and work order offers complete operational visibility, and manufacturing pedigree, including:

- Locating work order and calculating estimated time to completion
- Viewing product, employee, and workstation information
- Retrieving historical workstation, cycle time, and route

The solution was to add real-time sensing capabilities to that work cell structure, to create "smart" cells on a talking production floor, while letting the production simply flow. Endwave implemented the OMNITROL Smart Assembly Cell, which provides contactless tracking and real-time visibility onto the production line. The solution included Omnitrol's Work-In-Process application software running on Omnitrol's Edge Application Services Engine (EASE™), on the OMNITROL Appliance. The appliance functions as a server, database, intelligence layer, and integration layer. The appliance is both device- and platform-agnostic; any shop-floor device, be it RFID, RTLS, sensors, and so on, plugs into the appliance. In turn, the appliance integrates to any major enterprise system, be it SAP, Oracle, or the like.

Finally, the configuration included a database to track work-order pedigree, sales orders, materials, shipping, and so on. The system traced work through RFID-based hardware and software, which views the assembly cells as a chain of cells on which work orders are visible (see Figure 4.2).

Implementation took less than 30 days, from planning through testing. Endwave's Manufacturing Operations Department cooperated in outlining the production process, and identifying strategic locations for RFID-enabled smart shelves.

Although the implementation and testing lasted a month, visibility was immediate. Production managers could see work-order status on the Web interfaces and identify bottlenecks by the length of production

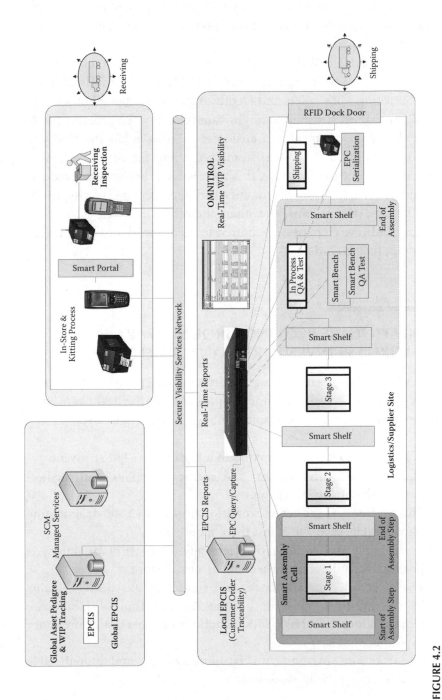

**FIGURE 4.2**

Lean and Wireless: cell-based manufacturing, with WIP visibility. (Courtesy Omnitrol Networks Inc.)

cycles. Each work order was trackable by both component and employee (pedigree), allowing the company to maintain records on components per product built.

Endwave expects the WIP Visibility solution to reduce costs comparable to one or more full-time employees per assembly line, by eliminating the labor of tracking work orders and identifying and managing bottlenecks. Production managers could communicate precise work-order status to both Endwave management and customers. Endwave has the ability to open that visibility data to its customers through the readily available, standards-based EPCIS interface; this too allows both the aerospace and DoD customers to look into the manufacturing process (as they required). The RFID tags, readers, bins, and smart shelves do the work of WIP visibility; no Endwave employee is required to fill out work orders or travelers. Instead, they fill orders.

---

## THE HEALTHCARE PROVING GROUND, PART 1: LEAN WIRELESS INVENTORY MANAGEMENT

If Lean Wireless is proven in a hospital, then, it is well proven indeed. Lean practices promote as little inventory as possible, but the healthcare environment defies every Lean principle. Among the challenges are:

- A hospital or clinic must maintain a high-value "buffer inventory" of pharmaceuticals, surgical and medical supplies, and medical devices, that it may never use.
- That inventory is subject to expiration (e.g., of a $2,500 drug-eluting stent, with its shelf-life of just three months).
- A stock-out at a catheter or electrophysiology (EP) laboratory is potentially fatal, versus a stock-out at a retail store.
- The inventories are regulated, and subject to unscheduled audits by the Joint Commission on the Accreditation of Healthcare Organizations (JCAHO), which checks stock levels and inventory expiration dates.
- The acute care environment is far more dynamic and unpredictable than is retail or manufacturing.

"The problem that healthcare experiences, along with everyone else, is loss; it doesn't matter if it is stolen or you can't find it," wrote clinical consultant and registered nurse Ann-Marie McDonough in *Cath Lab Digest*.[18] Loss of any kind, in a retail or manufacturing setting, contributes to the cost of goods sold (or services provided).

Put it all together, and inventory in clinical settings is difficult to manage. Because it might be pulled from a shelf (such as a cardiac stent, which must be "fitted" to a patient like shoes), it does not necessarily find its way back. And finally, the task of inventory is labor intensive, in a field in which clinical skills must be a priority.

"The sales rep has to do it," said John Wass, CEO of WaveMark. WaveMark provides real-time inventory management (RTIM) solutions for high-cost specialty products in the healthcare supply chain; this includes CIMS, the clinical inventory management system. Given the value of the inventory, it makes sense to use vendor-managed inventory, or VMI. But the clinical setting is too dynamic, said Wass. "There are always discrepancies that take a massive amount of time to resolve, from both the hospital and sales force." So vendor-managed inventory saves no time whatsoever; instead it wastes the time of *two* organizations. Wass believes that up to 40 percent of a medical device representative's time is wasted in administrative tasks such as these.

And in an acute care setting, the priority is treating a patient. Wass observes that missing inventory is often found in the laundry or trash.

## A Lean Wireless Answer

All of those challenges are enough to send a continuous improvement expert screaming over the hill, but Six Sigma Black Belt Lynda Wilson was undaunted. Wilson is the administrative project manager at Mercy Medical Center in Des Moines, Iowa. She was tasked with solving those inventory challenges at the Center's Cardiac Catheterization laboratory. Wilson implemented WaveMark CIMS, in a configuration that included:

- Sixteen RFID-enabled smart cabinets in procedure and store rooms
- RFID point of service (POS) readers in control rooms where documentation of cases is performed

The cabinets and POS report inventory status to a database, and make that status available through a Web interface. The systems are typically plug-and-play, with the cabinets powered by standard electrical outlets in the wall, and connecting into the clinic's network.

The point-of-service readers are installed by treatment tables, so that a nurse or clinician need only wave a tagged item by the reader; this records the use accurately, but spares the nurse or clinician having to record the entry. If the item is expired, or recalled, a red light and alarm alerts the technician. Finally, the smart cabinets record product entering or leaving the shelved inventory.[19] Neither the POS nor the smart cabinets require any key strokes or manual scans.

Nor, as is a fear with radio-frequency devices, do they interfere with electromagnetic and medical devices. The 13.56 MHz high-frequency used in WaveMark equipment is approved by the Food and Drug Administration (FDA)'s Center for Devices and Radiological Health (CDRH), and is restricted by the Federal Communications Commission for field strength. Finally, the equipment uses an industry standard ISO15693 RFID protocol, and contains read ranges to 6–10 inches, too short to interfere with medical equipment.

"It's been a wonderful tool in keeping track of expired product and consigned product," said Wilson. Whereas before the Center had to conduct a once-daily inventory of 2,200 items, the WaveMark system conducts an inventory 20,000 times yearly, or more than twice an hour. Rather than conduct the inventory, Center personnel spend perhaps 20 minutes per day tagging new inventory; or, as in the case with a few select vendors, the inventory arrives tagged.

The system returned more than 500 percent ROI in 18 months and enabled the clinic to reduce inventory levels by $376,587 (or 25 percent) in the first half of 2008, as well as reduce waste from expired products by 40 percent, from June 2007 to June 2008. And it saved 1.5 hours per weekday, or about 7.5 hours per week in conducting manual inventory.

## Is This Truly Lean—And How Does It Apply to Other Verticals?

A cardiac patient at first glance has nothing in common with a pallet of hair spray on a loading dock, or a Ford Mustang in an assembly plant. But look again. A clinic like Mercy Medical's Cath Lab has a distinct value stream. "Ultimately, we're here for the patient," said Wilson. Lower the cost of inventory and you will lower the cost of treatment. Free doctors and nurses from administrative tasks, "Then you have better patient care. The patient is always number one."

As with any enterprise, a clinic or hospital has distinct spending priorities. In production, the newest equipment has the highest priority, whereas in a clinic, any equipment used directly in improving patient care takes priority; compared to that equipment, the telephone systems and office equipment can be surprisingly arcane.

Thus any time spent in continuous improvement, or money spent on technology, must (1) cost justify itself and (2) ultimately serve the core mission (be it to manufacture an automobile or treat cardiac patients). In the current tough economic environment, said Wilson, "There are no capital dollars, no extra spend." Lean Wireless must pay for itself, several times over, if possible.

As a Six Sigma Black Belt, Wilson is focused upon results, but looks beyond Six Sigma tools. "We're limited on the time we have," said Wilson, "and you can't pull people out of cases or patient care to do a true DMAIC* project," which can take months, "whereas if you can hone in and use the skills of Lean to make it a quicker, less painful process, it makes it a success. So we use all the tools," be they Lean, Six Sigma, or Wireless.

What Wilson finds herself doing is drawing the connection to patient care and satisfaction. Healthcare professionals tend to shy from terms like Lean and Six Sigma, but are keenly interested in improving treatment and clinical results. "We're constantly making process improvements here at Mercy," said Wilson, and the hospital has seven steering committees that meet monthly or bimonthly. "But if you ask them 'how many Six Sigma or Lean projects have we done?' they'll say, 'None.'"

## Measurements and Metrics

There is no escaping the need for real-time information and key performance indicators; thus the Mercy Medical solution became a Six Sigma measurement tool. As Wilson put it, "The continuous improvement process needed improving," which measurement and reporting helped to do.

---

* The Six Sigma practice of define/measure/analyze/improve/control.

Mercy Medical's seven steering committees rely upon keen metrics. The committees may be patient- and treatment-focused, like the Cardiac Interventional Steering Committee of physicians, or more business operations focused, like an inventory steering committee, which involves administrative and finance personnel.

## The End-to-End Supply Chain

The healthcare field, as well, has become a proving ground for end-to-end supply chain traceability. As WaveMark CEO John Wass describes, "We find that the value of RFID—the real value—is to track expensive items that are critical to someone, all the way to the point of use, in a way that no other technology can. We focus on disposable and implantable products, from the point of manufacture to the point of implantation."

This is critical, and can be literally a matter of life or death, in the event of a product recall; this is why hospitals and clinics benefit from a central intelligence clearinghouse such as WaveMark, which stays current upon recalls and expiration dates. "That's a huge patient safety value," said Wass, "the ability to be certain that recalled product won't be used, because we can locate those at the hospitals." Here again, a technician is alerted by the WaveMark equipment with a red light and alarm indicator.

"So we literally go right to the point in the process where the product is used—the point of care. And by creating demand visibility, seeing what's needed at the moment it's needed, that's having a huge impact on overall productivity and efficiency of the value stream," which is of course, patient treatment. That, Wass believes, is "a very disruptive innovation in the healthcare field." The technology is disruptive to the current process in a positive way.

## Technology as a Tool in Healthcare

Wass and Wilson agree with the Lean credo that technology is not a solution; it is a tool. In the case of Mercy Medical Center, improved inventory visibility was the solution. "On other hand," said Wass, "if you can use innovative technology to reach a solution, wonderful!"

"I'm a huge advocate of RFID," said Wilson. "Knowing where things are in real-time is of value; if you have RFID technology and use it for tangible items, products or wheelchairs, and if you can avoid a loss by tracking items, then your costs are less, which is passed on," to first the healthcare provider, then to the patient.

In Chapter 3, we saw "simplification as a service," which is a powerful model for Lean Wireless providers such as WaveMark. The company's business model is an annual subscription, which eliminates both financial risk of upfront capital (once again, anathema to a healthcare provider), "And we insulate the hospitals from a technology list," said Wass. Lean Wireless technology is technically dense, but at its best, easily implemented and used. Mercy Medical required virtually no training or implementation downtime. Ultimately, said Wass, the WaveMark customers do not pay for RFID. "They pay us for accurate information."

## Pull-from-Demand

Moving from "push" to "pull-from-demand" is an ideal and as are all ideals, difficult to achieve. Enterprises of all kinds have evolved over time from (1) relying on historic demands to (2) complex forecasting, which factors historic demands with current conditions, to (3) responding to actual demand. No one pushes production from behind; rather, it is pulled by orders.

A common gripe in pull-from-demand retail is that it is impossible to connect the consumer to the manufacturer. But that's unnecessary. A retailer must connect the shelf to the manufacturer. "I think the battle for the consumer is increasingly going to be fought and won on the store shelf," said Procter & Gamble Product Supply Officer R. Keith Harrison, Jr. in *RFID and Beyond*.[20]

If we are to move toward vendor-managed inventory, or VMI—which takes the burden of managing inventory off the retailer and moves it to the manufacturer, who uses it for pull-from-demand manufacturing—then these connections have to be the norm.

Moreover, the systems must be completely automated, so that the systems generate a work order or purchase order rather than wait for someone to do it. But they are proven across six suites in Mercy Medical. And if that is not a large enough example, then, they are proven across the wildly successful American Apparel retail stores.

## THE HEALTHCARE PROVING GROUND, PART 2: RTLS ADDS VALUE

As we described in the previous section, the first priority in a hospital or clinic is patient care; a hospital will upgrade its diagnostic equipment long before it upgrades nonclinical systems (such as telephones).

Real-time location systems are on the rise in hospitals, because they add value to the core mission of patient care, well in excess of their cost. But in healthcare, the value is measured differently from other industries:

$$\text{Healthcare Value} = \text{Quality of Care} \times (\text{Revenue/Cost}) - \text{Risk}$$

This is not easy mathematics: revenue, cost, and risk can be measured in dollars, but quality of care cannot. That is a more subjective measure, based upon patient satisfaction and other factors. Still, RTLS can be used to directly increase all of these variables.

### Controlling Cost and Risk

Thornton Hospital at the University of California San Diego (UCSD) Medical Center selected the Awarepoint RTLS (which it calls a real-time awareness system, or RTAS) to fulfill those requirements. In 2005, Thornton Hospital's Director of Preoperative Services, Tom Hamelin, sought an RFID solution to address the challenges of:

- Lowering equipment rental costs
- Reducing staff time spent searching for equipment
- Minimizing equipment theft and loss
- Reducing equipment inventory requirements
- Improving equipment maintenance process
- Improving responsiveness to JCAHO and FDA requirements

Hamelin selected Awarepoint because it is 100 percent wireless, and the company could provide nondisruptive installation, low cost to trial (owing to no construction costs), and a fully managed service model allowing the hospital to easily scale up the system on a per-month basis. Thus, the risk of disruption to patients and risk of contamination that accompanies construction, was low. Full implementation took under three weeks at this 119-bed, 238,792-square-foot general medical–surgical facility, and has since been expanded to the medical center's larger Hillcrest campus and Moore's Cancer Center.

In just four months following the Thornton implementation, the hospital reduced its rental of IV pumps from $8,000 per month to $2,000 per month, simply by improving its ability to locate the pumps. As of this writing, Thornton has maintained this cost reduction for more than 18 months, a cumulative savings of nearly $110,000 thus far. In terms of patient safety and compliance, the hospital found itself able to react rapidly to an FDA "Urgent Medical Device Recall," identifying 200 recalled IV pumps in under 48 hours; such recalls would have taken several weeks and dozens of hours of staff labor in the past. It is a very short leap between recall response and contagion response. This both reduces the cost of care, and reduces the risk.

The University of California San Francisco (UCSF) Medical Center similarly implemented Awarepoint RTAS at its 1.2-million-square-foot academic facility. The Center freed up 1,600 hours per year in operating-room staff time, simply in locating equipment. Far from being just an efficiency measure, delays in locating critical equipment "show stoppers" can delay surgery, and as a result, affect both patient safety and the hospital's topline revenue. As a further example of efficiency, UCSF estimates that it has reduced finding a lost asset from a 30–45-minute task to a "3-second task." Prior to RTLS, cash-strapped hospitals have been tasked with purchasing redundant equipment to ensure adequate supply is on hand for patient care. With Awarepoint, UCSF avoided a planned purchase of additional transport monitors and intubation equipment, freeing $248,000 to be saved or spent on new technology. Thus, UCSF moved all variables in the healthcare value equation, by increasing revenue, reducing cost, reducing risk, and increasing quality.

This enterprise awareness is, again, what is required in both rapid response to infection, and in identifying exposures.

## DEMOCRACY MARCHES ON

"Swell," the reader might say, after all these success stories. "But I don't handle baggage, I don't arrest felons, I'm not running for election, and

I don't treat patients. I make toys. Is this stuff proven in toys?" Good question, and there are three answers.

First: undoubtedly, yes, it is proven in toys (or in manufacturing yachts, or in machine shops, or on college campuses, in whatever type of enterprise you work). We sifted through hundreds of studies, to find the most illustrative.

Second: it is useful to find both a Lean consultant and Wireless provider with success in your vertical. As we pointed out in Chapter 3, "The Lean Wireless Missions," the proof in Wireless is in the résumé. Certainly, a Wireless provider who has installed a system at Boeing knows the ins and outs of aerospace and defense.

Third, and this may be the most important point: don't limit yourself to what is proven in your vertical, particularly if your vertical is manufacturing, which has always been insular and self-referential about its technology. In the Wireless/Web 2.0 world, the best practices can come from anywhere, and anyone. They did not call it *muda*, but Mercy Medical, Jackson Hospital, and UCSD each used RFID or RTLS to reduce waste and turn the money back into superior patient care. They did not call it a value stream, or even a supply chain, but Alameda County, California used RFID to vastly improve the flow of its election-night process, and eliminated hours of wasted labor. Finally, the Thames Valley police force may never have heard the Lean adage, "Drill, baby, drill." But using Wireless, its officers spend 30 percent more time in the field, actively enforcing the law.

Used well, Wireless provides Lean benefits wherever it is used.

## ENDNOTES

1. Knowledge@Wharton. 2008. "Why an Economic Crisis Could Be the Right Time for Companies to Engage in 'Disruptive Innovation.'" http://knowledge.wharton.upenn.edu/article.cfm?articleid=2086 (accessed January 9, 2009).
2. Morgenroth, D. 2008. Reality check: RFID unlocks the potential one step at a time, *The RF Edge*, 3 November, http://www.rfid-world.com/features/supply-chain/212000291
3. Morgenroth, D. 2008. Reality check: RFID unlocks the potential one step at a time, *The RF Edge*, 3 November, http://www.rfid-world.com/features/supply-chain/212000291
4. Morgenroth, D. 2008. Reality check: RFID unlocks the potential one step at a time, *The RF Edge*, 3 November, http://www.rfid-world.com/features/supply-chain/212000291
5. Wagner, M. 2008. Obama election ushering in first Internet presidency, *Information Week*, November 10, 17.

6. Wagner, M. 2008. Obama election ushering in first Internet presidency, *Information Week*, November 10, 17.
7. Wagner, M. 2008. Obama election ushering in first Internet presidency, *Information Week*, November 10, 17.
8. Wagner, M. 2008. Obama election ushering in first Internet presidency, *Information Week*, November 10, 17.
9. Wagner, M. 2008. Obama election ushering in first Internet presidency, *Information Week*, November 10, 17.
10. Zeleny, J. 2008. Lose the BlackBerry? Yes, he can. Maybe, *The New York Times*, 15 November, L1.
11. Mears, B. 2009. "Court Orders White House to Preserve Emails," CNN Politics.com, 14 January (CNN.com.).
12. Research in Motion. 2007. "Cape Breton police force leads by example with a BlackBerry solution." Xwave.com. http://www.xwave.com/files/credentials/CapeBreton_lob_final.pdf (09 January 2009).
13. Rutanen, T. 2008. *Asset and People Tracking with Wi-Fi RTLS*. Ekahau, Inc. Saratoga, CA.
14. Saksena, R. and D. Orain, 2009. *Global track and trace at the edge*. Presentation to authors, March 23.
15. Rutanen, T. 2008. *Asset and People Tracking with Wi-Fi RTLS*. Ekahau, Inc. Saratoga, CA.
16. Rutanen, T. 2008. *Asset and People Tracking with Wi-Fi RTLS*. Ekahau, Inc. Saratoga, CA.
17. Interview with Tuomo Rutanen, Ekahau, Inc.
18. McDonough, A.M. 2008. RFID in the cath lab: Real-time inventory management supports patient safety. *Cath Lab Digest* (10 October 08): http://www.cathlabdigest.com/articles/RFID-Cath-Lab-Real-Time-Inventory-Management-Supports-Patient-Safety
19. McDonough, A.M. 2008. RFID in the cath lab: Real-time inventory management supports patient safety. *Cath Lab Digest* (10 October 08): http://www.cathlabdigest.com/articles/RFID-Cath-Lab-Real-Time-Inventory-Management-Supports-Patient-Safety
20. Heinrich, C. 2005. *RFID and Beyond*. Indianapolis: Wiley.

# Afterword (and Forward): Lean Wireless 2015

There are many bromides applicable here—too much of a good thing, tiger by the tail, as you sow so shall you reap. The point is that too often man becomes clever instead of becoming wise, he becomes inventive but not thoughtful.

**Rod Serling, speaking about automation. From the 1964 episode of *The Twilight Zone*, "The Brain Center at Whipples."**

As far back as *Frankenstein*, nonscientists have warned that our technological hubris would bite us back.

In the episode of *The Twilight Zone* we quoted, a store owner named Whipple automates his store, and is then fired by a computer. In *The Island of Dr. Moreau*, the doctor is slaughtered by a hyena/swine monster that he creates in his lab. In *Planet of the Apes* we are enslaved by the apes we once enslaved, and in the *Terminator* films, enslaved by the machines that we created. According to this formula, the Wireless technology that we use in our laptops and PDAs, automobiles, and access cards will somehow do the same.

Interestingly, *The Terminator* turns up frequently in relation to Wireless, the Internet, and robotics. "Call it 'Terminator Vision,'" wrote the BBC, about the rise of "augmented reality" in mobile and smartphones. Point your phone down the street, see where the public transportation is; houses for sale; a barber shop, if you want one, or a flower shop; a short video about the location. "Eventually, it seems possible that mobile phones might play the role of a kind of supplementary brain," Japanese tech journalist Toshinao Sasaki enthused to the BBC.[1] Less ebullient was the *Times of London* headline, "Military's killer robots must learn warrior code." The *Times* detailed a gloomy report, "the first serious report of its kind" on robot ethics, compiled by California State Polytechnic University for the Office of Naval Research. The report specifically named the *Terminator* films as a model of military robots gone bad. Dr. Patrick Lin of Cal State Polytech warned that "There is a common misconception that robots will do only what we have programmed them to do... ." But programming can be contradictory, faulty, and hacked. "We are going to need a warrior code" of battlefield ethics, said Lin; he specifically named Isaac Asimov's three laws of robotics, which we detailed earlier in Chapter 3.[2]

Author Kevin Kelly believes the fault is not in our machines, but in ourselves. Kelly argues that technology should be reduced as much as possible because it is "contrary to nature, and/or to humanity... . Technology erodes human character. It separates us from nature, which diminishes our natural self. Out of touch with nature, we behave selfishly, stupidly."[3] Kelly is no technophobe; he is the author of *Out of Control: The New Biology of Machines, Social Systems, and the Economic World*, is also the "Senior Maverick" at *Wired* magazine, and publishes the "Cool Tools" Web site. Kelly does not separate the socio- from the economic; neither have we in *Thin Air* because they are inseparable.

## THE REALITY

Unlike "Whipple's" in the *Twilight Zone* episode, American Apparel did not implement RFID and sack all the humans; it opened more stores and hired more humans.

There are plenty of dreadful stories about individuals using the Internet to harass one another; but there has yet to be a significant tale about Wireless technology or the Internet making their own decisions and biting us back. Nor have they proven particularly evil. Certainly, a few hysterics have decried RFID as the "Mark of the Beast," the badge of the biblical Antichrist, without which as the Scriptures go, "None shall buy or trade"; but bar codes have been similarly decried, as was the Internet, as were SIC codes, and so on. Still, after 60 years of automation with humans in charge, there are those who believe we will lose our jobs and surrender control to computers and machines.

It will never happen. "The interesting thing about most information technology, if not all IT today, is that it is *still very dependent on human beings*," said Kevin Ashton, the fellow who coined the term, "The Internet of Things." "It's human-entered, and processed by human beings, with very few exceptions. As excited as we get with the digital revolution, we manually scan bar codes, and manually take photos and email them to another. There tends to be a human being at the beginning and the end of the process, and that's today's Internet." We do not surrender control by automating decisions, or using RFID to track inventory; we extend our control.

We selectively surrender our control by using Expedia to find flights and Priceline to find hotels for us. Still, we surrender that control not to an algorithm on a computer, but to a company and individuals who have earned our trust.

We will still be in control in the highly wireless world of 2015. It will be a very different world, one in which workers with Web-based immersive graphic environments will appear to be playing.

We will surrender privacy and work a good deal harder but, overall, we will be pleased to do so. (Rather, the next generation will be; we won't, necessarily.) The next generation will be able to spontaneously acquire any information or content, wherever it is; the Internet will be as ubiquitous, available, and invisible as radio once was. As futurist Glen Hiemstra described in 2006, "Information and global communications will be the sea that all humans swim in."[4]

We will carry our computing power in our pockets or in a badge or a pair of glasses. Just as a cell phone signal switches between towers, Internet connectivity will switch between nodes to provide uninterrupted access, at the breakfast table, walking to the train, in the office, and at the gym after work. We will likely move from the small screens of our PMAs to a full-sized PC at work, without interruption.

These tremendous gains in efficiency and productivity will eliminate scut work, but will mean higher expectations from individual workers. The automation that caught hold in the 1950s did not fulfill the promise of the four-day work week and one-month paid vacation, and neither will the efficiency gains of the early twenty-first century.

What are now called "Lean" and "continuous improvement" practices will be deeply entrenched and largely the norm across verticals. And wherever possible, the mechanics of Lean will have gone wireless. They hone workers' individual value streams, to the point at which a welder is either welding or at lunch; and pharmacists are dispensing medicines and advice, without the burden of paperwork and inventory checks. Drivers will drive, and well; pilots will fly, undistracted; and production operators will run machines, multiple machines.

Continuous improvement experts must involve themselves in electronic value streams. In the information age, information creates its own forms of waste. The old Lean aphorism, "Drill, baby, drill" makes sense when the tool is a drill; but the tools in the information age, chiefly the Internet, represent far more fun and risk and waste than a drill. Besides which, it is obvious when a worker is drilling, less so when he is text messaging.

Business will likely surrender control of workers' work habits, and focus upon results, from an individual worker or wireless device. Perhaps continuous improvement practitioners will lean that way as well, focusing on ways to measure results (as Six Sigma does) rather than to doggedly eliminate waste.

Anything that can be wireless will in time be wireless; it is too cost effective and has too much momentum to stop, but it cannot barrel on without forethought. Business needs Lean and Six Sigma experts to do what they do best: identify opportunities for both incremental and grand-scale improvements, and eliminate the waste.

We hope that *Thin Air* helps to foster a new cooperation in business; that it serves to temper the suspicion of wireless technology, while checking its runaway speed. Used well, Wireless does remarkable things for industry and individuals. We call upon the technology providers and continuous improvement experts to see that it is used well.

## ENDNOTES

1. Fitzpatrick, M. 2009. Mobile phones get cyborg vision. *BBC News*. http://news.bbc. co.uk/2/hi/technology/8193951.stm (accessed September 1).
2. Lewis, L. 2009. "Military's Killer Robots Must Learn Warrior Code." *The TimesOnline*. http://technology.timesonline.co.uk/tol/news/tech_and_web/article5741334.ece (accessed September 1).
3. Kelly, K. 2009. "4 Arguments Against Technology." http://blogs.harvardbusiness.org/ now-new-next/2009/04/4-arguments-against-technology.html (accessed September 1).
4. Hiemstra, G. 2006. "Eleven Events, Trends and Developments That Will Change Your Life." futurist.com.

# Appendix A: Lean Glossary*

Lean is an integrated approach to designing and improving work toward a customer-focused ideal state through the engagement of all people aligned by common principles and practices.

## LEAN PRINCIPLES

Lean principles guide our thinking, our decision making, and the way we see things. It is important in a Lean environment to have shared thinking, and these lean principles are the foundation for that shared thinking. Missing this element is the cause of many lean failures.

1. Directly observe work as activities, connections, and flows.
2. Systematic waste elimination.
3. Systematic problem solving.
4. Establish high agreement of both what and how.
5. Create a learning organization.

## LEAN RULES

Lean rules are a guide, helping us as we design, operate, and improve our organizations. They help us understand in which direction to go when looking at our processes. These rules were originally articulated through the research of H. Kent Bowen and Steven Spears but are modified for ease of use and memory.

1. Structure every activity.
2. Clearly connect every customer–supplier.

---

* Courtesy of the Lean Learning Center, Inc., www.leanlearningcenter.com

3. Specify and simplify every flow path.
4. Improve through experimentation.

---

## LEAN CONCEPTS

Lean concepts is not necessarily a tool that we would implement but a state or condition for which we would strive. For example, a "balanced diet" is a concept to strive for but a specific diet name or structure would be a tool to help us achieve that concept.

**Jidoka:** Also referred to as autonomation, it is adding the human element of being able to identify problems and either stop for correction or self-correct before moving on to the next step.

**One-piece flow/continuous flow:** The ideal state for any process is to move away from traditional batching of work, whether material or information, and flow work continuously, one element at a time. This reduces many types of waste, particularly inventory.

**PDCA:** Plan-Do-Check-Act means that whether solving a problem or building a plan everyone should follow this process to ensure learning and success toward the goal.

**Pull:** In order to improve continuous flow and reduce the waste of overproduction, processes should "pull" what they need from the previous step in the process, and only that triggers new actions.

**Value-added:** Value-adding tasks are only those tasks that (1) the customer is willing to pay for, (2) transform the product or service, and (3) are done right the first time.

**Visual management:** Visual management is simultaneously a tool and a concept. The ideal state is that all employees, operators, and management should be able to manage every aspect of the process at-a-glance using visual data, signals, and guides.

**Waste elimination:** Eliminating waste from the process is the goal of many lean tools and should be an ongoing effort in itself. This comes in the form of the seven types of waste: overproduction, waiting, inventory, overprocessing, motion, transportation, and defects.

## LEAN TOOLS

Lean tools are proven practices that help us move closer to our ideal state, help us apply the four Lean rules, and are consistent with Lean principles. This section does not articulate the how of these tools; it simply defines several, but not all, of them.

**5Ss:** 5Ss are adapted from five Japanese words that start with "s" but have been rewritten as Sift, Sweep, Sort, Sanitize, and Sustain. It helps us organize what we need and eliminate what we don't, allowing us to identify problems quickly.

**5 Ws:** The five Ws (whys) is a method of solving a problem by asking why the problem occurred, and then why did that cause occur, and so on until you get to the root cause of a problem.

**Andon:** The Andon cord is the ability for an operator to pull a cord that triggers a horn and light which tells the team leader or supervisor that he needs help or support. Once provided, the team leader can pull the cord to keep production moving.

**Cell:** A cell is an arrangement of work (machines, people, method, and material) so that processing steps are sequentially done in next-door steps one at a time in order to improve efficiency, reduce waste, and improve communication.

**Error proofing:** Error proofing is also known as poka-yoke or mistake proofing. It involves the redesign of equipment or processes to prevent problems from occurring or allowing defective product to move on.

**Hoshin kanri:** A strategic planning process to establish high agreement and align people in a common direction with agreed-upon methods to improve a process.

**Kaizen:** Kaizen is a structured process to engage those closest to the process to improve both the effectiveness and efficiency of the process. Its goals are often to remove waste and add standardization.

**Kanban:** Kanban, often in the form of cards, are a signal that a downstream or customer process can use to request a specific amount of a specific part from the upstream, or supply, process.

**Preventive maintenance:** Simplifying and structuring maintenance activities to prevent problems rather than react to them can increase capacity and improve continuous flow.

**Scoreboards:** Scoreboards are visual management of safety, quality, delivery, and cost metrics including analysis and action plans used to help shop floor teams manage their own process.

**Setup reduction:** The time it takes to change equipment over from one product to the next is a major barrier to continuous flow, and setup reduction seeks the reduction or elimination of that time. This is also known as SMED, or single-minute exchange of dies.

**Six Sigma:** Six Sigma is a method and a set of tools to reduce variation in processes, particularly quality, using mostly statistical tools. Its primary method is DMAIC: Define, Measure, Analyze, Improve, and Control.

**SWIs:** SWIs, or standard work instructions, are a visual method of structuring every job, providing easy access to key information for operators and allowing for continuous improvement.

**Value stream mapping:** This structured process helps managers understand the flow of both material and information through their operation and develop plans to move them closer to the ideal state.

# Appendix B:
# Wireless Glossary*

## A

**Active tag:** RFID tag that uses batteries as a partial or complete source of power to boost the effective operating range of the tag and to offer additional features over passive tags, such as temperature sensing.

**Active transponder:** See Active tag.

**Agile reader:** A generic term that usually refers to an RFID reader that can read tags operating at different frequencies or using different methods of communication between the tags and readers.

**AIM:** The industry association for Automatic Identification and Mobility.

**Air interface:** The conductor-free medium, usually air, between a transponder and a reader/interrogator through which data communication is achieved by means of a modulated inductive or propagated electromagnetic field.

**Airsource:** The automation of a routine task using a wireless device, akin to outsourcing in that the goal is to allow workers to focus upon value-add activity.

**Alignment:** The orientation of a transponder relative to the reader/interrogator antenna. Alignment can influence the degree of coupling between transponder and reader, separation being a further influence.

**Alphanumeric:** Data comprising both alphabetical and numeric characters, for example, A1234C9 as an alphanumeric string. The term is often used to include other printable characters such as punctuation marks.

**Amplitude:** The maximum absolute value of a periodic curve measured along its vertical axis (the height of a wave, in layman's terms).

---

* Compiled with the cooperation of AIM Global, the industry association for Automatic Identification and Mobility, www.aimglobal.org, Zebra Technologies (www.zebra.com), and Industry Wizards (www.industrywizards.com).

**Amplitude modulation:** Changing the amplitude of a radio wave. A higher wave is interpreted as a one and a normal wave is interpreted as a zero. By changing the wave, the RFID tag can communicate a string of binary digits to the reader. Computers can interpret these digits as digital information. The method of changing the amplitude is known as amplitude shift keying, or ASK.

**ANSI (American National Standards Institute):** An American technical standards body and the representative of the United States to the International Organization for Standardization.

**Antenna:** A device that conducts electromagnetic energy. In RFID, an antenna radiates energy in the radio frequency spectrum to and from the RFID tag.

**Auto-ID Center:** The nonprofit organization which led the development of a global network for tracking goods academia, that pioneered the development of an Internet-like infrastructure for using RFID to track goods globally.

**Automatic Identification (or Auto-ID):** This broad term covers methods of collecting data and entering it directly into computer systems without human involvement. Technologies normally considered part of auto-ID include bar codes, biometrics, RFID, and voice recognition.

**Awake:** The condition of a transponder when it is able to respond to interrogation.

# B

**Bandwidth:** The range or band of frequencies, defined within the electromagnetic spectrum, that a system is capable of receiving or delivering.

**Bar code:** A standard method of identifying the manufacturer and product category of a particular item. The bar code was adopted in the 1970s because the bars were easier for machines to read than optical characters. Bar codes' main drawbacks are they don't identify unique items and scanners have to have line of sight to read them.

**Batch reading:** The process or capability of a radio frequency identification reader/interrogator to read a number of transponders present within the system's interrogation zone at the same time. It is an alternative term for multiple reading.

**Battery-assisted tag:** RFID tags with batteries, but they communicate using the same backscatter technique as passive tags (tags with no battery). They use the battery to run the circuitry on the microchip and sometimes an onboard sensor. They have a longer read range than a regular passive tag because all of the energy gathered from the reader can be reflected back to the reader. They are sometimes called "semi-passive RFID tags."

# C

**Capacity – Channel:** A measure of the transmission capability of a communication channel expressed in bits and related to channel bandwidth and signal-to-noise ratio by the Shannon equation; Capacity, $C = B \log^2 (1 + S/N)$, where B is the bandwidth and S/N the signal-to-noise ratio. Compare Capacity – Data.

**Capacity – Data:** A measure of the data, expressed in bits or bytes, that can be stored in a transponder. The measure may relate simply to the bits that are accessible to the user or to the total assembly of bits, including data identifier and error control bits. Compare Capacity – Channel.

**Capture field/area/zone (also Interrogation zone/area/volume):** The region of the electromagnetic field determined by the reader/interrogator antenna, in which the transponders are signaled to deliver a response.

**Card operating system:** The software program stored in the smart card IC, which manages the basic functions of the card, such as communication with the terminal, security management, and data management in the smart card file system.

**Carrier:** The abbreviated term for carrier frequency.

**Carrier frequency:** Used to carry data by appropriate modulation of the carrier waveform, typically in a radio frequency identification system, by amplitude shift keying (ASK), frequency shift keying (FSK), phase shift keying (PSK), or associated variants. See also Tolerance.

**Channel:** The medium or medium associated allocation, such as carrier frequency, for electronic communication.

**Character set:** A set of characters assembled to satisfy a general or application requirement.

**Chip:** In data communication terms, this is the smallest duration of a pseudo-random code sequence used in spread spectrum communication systems.

**Chipless RFID tag:** An RFID tag that doesn't depend on a silicon microchip. Some chipless tags use plastic or conductive polymers instead of silicon-based microchips. Other chipless tags use materials that reflect back a portion of the radio waves beamed at them. A computer takes a snapshot of the waves beamed back and uses it as a fingerprint to identify the object with the tag. Companies are experimenting with embedding RF reflecting fibers in paper to prevent unauthorized photocopying of certain documents. Chipless tags that use embedded fibers have one drawback for supply chain uses—only one tag can be read at a time.

**Closed systems (closed loop systems):** Within the context of radio frequency identification, they are systems in which data handling, including capture, storage, and communication, is under the control of the organization to which the system belongs. Compare with Open systems.

**Commissioning a tag:** This term is sometimes used to refer to the process of writing a serial number to a tag (or programming a tag) and associating that number with the product it is put on in a database.

**Contactless smart card:** An awkward name for a credit card or loyalty card that contains an RFID chip to transmit information to a reader without having to be swiped through a reader. Such cards can speed checkout, providing consumers with more convenience.

**Continuous reporting:** A mode of reader/interrogator operation wherein the identification of a transponder is reported or communicated continuously while the transponder remains within the interrogation field. See also In-field reporting.

**Corruption-data:** In data terms, this encompasses the manifestations of errors within a transmitted data stream due to noise, interference, or distortion.

# D

**Data:** The representations, in the form of numbers and characters, for example, to which meaning may be ascribed.

**Data field:** A defined area of memory assigned to a particular item or items of data.

**Data field protection:** The facility to control access to and operations upon items or fields of data stored within the transponder.

**Data rate (Data transfer rate):** In a radio frequency identification system, this is the rate at which data is communicated between the transponder and the reader/interrogator, expressed in baud, bits or bytes.

**Data transfer:** The process of transferring data from a data holding source to a destination.

**Data transfer rate:** The rate at which data is transferred between the reader and a tag, generally measured in bits per second (bps).

**Digital certificate:** A digital message that contains the identity of a company or organization, its public key combined, and a signature of this data from a certificate authority (Trust Center) proving the correctness of this data.

**Dipole (antenna):** A fundamental form of antenna, comprising a single conductor of length approximately equal to half the wavelength of the carrier wave. It provides the basis for a range of other more complex antenna forms.

## E

**EAN International:** The international organization responsible for administering bar-code standards.

**Electromagnetic energy:** A process of transferring modulated data or energy from one system component to another, reader to transponder, for example, by means of an electromagnetic field.

**Electromagnetic field:** The spatial and temporal manifestation of an electromagnetic source in which magnetic and electric components of intensity can be distinguished and plotted as contours, like contour lines on a map, the planes of the electric and magnetic contours being at right angles to each other. Where the source is varying in time so too the field components vary with time. Where the source launches an electromagnetic wave the field may be considered to be propagating.

**Electromagnetic spectrum:** The range or continuum of electromagnetic radiation, characterized in terms of frequency or wavelength.

**Electromagnetic wave:** A sinusoidal wave in which electric E and magnetic H components or vectors can be distinguished at right angles to each other, and propagating in a direction that is at right angles to both the E and H vectors. The energy contained within the wave also propagates in the direction at right angles to the E and H vectors. The power delivered in the wave is the vector product of E and H (Poynting vector).

**Electronic article surveillance:** This is accomplished by simple electronic tags that can be turned on or off. When an item is purchased (or borrowed from a library), the tag is turned off. When someone passes a gate area holding an item with a tag that hasn't been turned off, an alarm sounds. EAS tags are embedded in the packaging of most pharmaceuticals. They can be RF-based, or acoustomagnetic.

**Electronic data interchange (EDI):** The communication of a data message, or messages, automatically between computers or information management systems, usually for the purposes of business transactions.

**Electronic data transfer (EDT):** The transfer of data by electronic communication means from one data handling system to another.

**Electronic label:** An alternative colloquial term for a transponder.

**Electronic Product Code (EPC):** An identification standard created by the Auto-ID Center that provides additional information to existing bar codes. The EPC can identify manufacturers, product categories, and individual items. See also Auto-ID center and Bar code.

**EPC Discovery Service:** A service from the EPCglobal Network that allows companies to search for every reader that has read a particular EPC tag. See also EPCglobal.

**EPC Information Service:** A network infrastructure of the EPC Network that allows companies to store EPC data in secure databases on the Web, and enables companies to set the level of information access for different types of organizations, from supply chain providers to manufacturers, to everyone.

**EPCglobal:** A non-profit organization with the mission of commercializing EPC technology. The Uniform Code Council and EAN International, which set and maintain bar-code standards in North America and internationally, set up EPCglobal.

**EPCglobal Network (or EPC Network):** The Internet-based technologies and services designed for EPCs.

**ePedigree or Electronic pedigree:** A secure file that stores data about each move a product makes through the supply chain. Pedigrees can help to reduce counterfeiting of drugs and other products.

**Encryption:** A means of securing data, often applied to a plain or clear text, by converting it to a form that is unintelligible in the absence of an appropriate decryption key.

**Environmental parameters:** Parameters, such as temperature, pressure, humidity, or noise that can have a bearing or impact upon system performance.

**Error:** In digital data terms, error is a result of capture, storage, processing, or communication of data in which a bit or bits assume the wrong values, or bits are missing from a data stream.

**Error burst:** A group of bits in which two successive erroneous bits are always separated by less than a given number of correct bits.

**Error control:** A collective term to accommodate error detection and correction schemes applied to handle errors arising within a data capture or handling system. See also Redundancy.

**Error detection:** A scheme or action to determine the presence of errors in a data stream.

**Error rate:** The number of errors divided by the number of transactions.

**European Article Numbering (EAN):** The bar-code standard used throughout Europe, Asia, and South America, administered by EAN International.

**Event data:** Information related to a transaction or incident with significance to the business. If a tag on a pallet is read as the pallet leaves a dock door, an event is recorded (the pallet was shipped). If a reader reads a tag on a pallet in a storage bay 100 times per minute but the pallet never moves, data is generated, but there is no event.

**F**

**Factory programming:** The entering of data into a transponder as part of the manufacturing process, resulting in a read-only tag. Compare Field programming.

**Far field communication:** The region of an electromagnetic radiation field at a distance from the antenna in which the field distribution is unaffected by the antenna structure and the wave propagates as a plane wave. Compare Near field communication.

**Field programming:** The programming of information into a tag after shipment by the manufacturer, either by an OEM customer or end user. Field programming is often related to the tag's target application.

**File:** A set of data stored within a computer, portable data terminal, or information management system.

**Firmware:** The coded instructions that are stored permanently in read-only memory. When upgrading a reader to read a new protocol, the firmware usually has to be changed. Some newer readers can be upgraded remotely over a network.

**Flat panel antenna:** Flat antennas that are generally made of metal plate or foil and embedded in a label or other material.

**Form factor:** The packaging in which a transponder can be put. These include thermal transfer labels, plastic cards, key fobs, and so on.

**Frequency:** The number of cycles a periodic signal executes in unit time. Usually expressed in hertz (cycles per second) or appropriate weighted units such as kilohertz (kHz), megahertz (MHz), and gigahertz (GHz).

# G

**Global Commerce Initiative:** A user group founded in October 1999 by manufacturers, retailers, and trade industry associations, to improve the performance of the international supply chain for consumer goods through the collaborative development and endorsement of recommended voluntary standards and best practices. Its charter is to drive the implementation of EAN.UCC standards and best practices, including use of EPC.

**Global data synchronization:** This term generally refers to the process of ensuring that a manufacturer's master files with product information match those of retailers. GDS is an important prerequisite to deploying RFID in open supply chains because companies need to ensure that RFID serial numbers refer to the right product information in a database.

**Global location number:** A numbering scheme created by EAN International and the Uniform Code Council as a means to identify virtually limitless numbers of legal entities, trading parties, and locations to support the requirements of electronic commerce (B2B and B2C). Parties and locations that can be identified with GLNs include functional entities (e.g., a purchasing, accounting, or returns department), physical entities (e.g., a particular room in a building, warehouse, loading dock, or delivery point), and legal entities or trading partners (e.g., buyers, sellers, whole companies, subsidiaries or divisions such as suppliers, customers, financial services companies, or freight forwarders).

**Global Positioning System (GPS):** Developed for and managed by the United States military, GPS is a satellite navigation system. It consists of 24 satellites above the earth. They transmit radio signals to receivers placed on ships, trucks, or other large assets that need to be tracked. The receivers compute longitude, latitude, and velocity by calculating the difference in the time signals received from four different satellites. Some companies are integrating RFID and GPS systems to track assets in transit.

**Global System for Mobiles (GSM):** The digital cellular telephone system widely used in Europe, Asia, and Australia.

**Global Trade Item Number (GTIN):** A standardized system of identifying products and services created by the Uniform Code Council and EAN International. Product identification numbers, such as EAN/UCC-8, UCC-12, EAN/UCC-13, and EAN/UCC-14, are based on the GTIN.

# H

**Handshaking:** A protocol or sequence of signals for controlling the flow of data between devices, which can be hardware implemented or software implemented.

**Harmonics:** Multiples of a principal frequency, invariably exhibiting lower amplitudes, harmonics can be generated as a result of circuit nonlinearities associated with radio transmissions resulting in harmonic distortion.

**High frequency:** Generally considered to be from 3 MHz to 30 MHz. HF RFID tags typically operate at 13.56 MHz. They can be read from less than three feet away and transmit data faster than low-frequency tags. But they consume more power than low-frequency tags.

**High-frequency tags:** RFID systems that operate at 13.56 MHz with a typical maximum read range of up to 3 feet (1 meter).

**Host system:** A computer on a network, which provides services to users or other computers on that network is called a host system.

**Hybrid card:** A smart card that has both a contactless IC and a contact IC. Unlike a dual interface card, a hybrid card acts as two separate cards.


# I

**ID filter:** A software facility that compares a newly read identification (ID) with those within a database or set, with a view to establishing a match.

**iGemba:** A virtual visit to a work space, using a Web-based application.

**iKaizen:** The practice of using social media or online collaborative technology to engage those closest to a process in removing waste and standardizing the process.

**Impact:** An impact is any influence upon a system, environmental or otherwise, that can influence its operational performance.

**iMuda:** The waste generated by a Web-based or Wireless technology.

**In-field reporting:** A mode of operation in which a reader/interrogator reports a transponder ID on entering the interrogation zone and then refrains from any further reports until a prescribed interval of time has elapsed. See also Out-of-field reporting.

**In-use programming:** The ability to read from and write to a transponder while it is attached to the object or item for which it is being used. Compare Factory programming, Field programming.

**Incorrect read:** The failure to read correctly all or part of the dataset intended to be retrieved from a transponder during a read or interrogation process. Alternative term for Misread.

**Inductive coupling:** A process of transferring modulated data or energy from one system component to another, reader to transponder, for example, by means of a varying magnetic field.

**Inlay:** An inlay is RFID hardware mounted on label material. Inlays provide the RFID portion of "smart labels." See also Smart label.

**Intelligent reader:** This generic term is sometimes used to describe a reader that has the ability to filter data, execute commands, and generally perform functions similar to a personal computer.

**Interchangeability:** This comprises the condition that exists between devices or systems that exhibit equivalent functionality, interface features, and performance to allow one to be exchanged for another, without alteration, and achieve the same operational service. An alternative term for compatibility. Compare Interoperability.

**Interface:** An interface is a physical or electrical interconnection between communicating devices. See also RS232, RS422, and RS485.

**Interference:** Unwanted electromagnetic signals, where encountered within the environment of a radio frequency identification system, cause disturbance in its normal operation, possibly resulting in bit errors and degrading system performance and are termed interference.

**International Organization for Standardization (ISO):** The ISO is a nongovernmental organization made up of the national standards institutes of 146 countries. Each member country has one representative and the organization maintains a Central Secretariat in Geneva, Switzerland that coordinates the system.

**Internet of Things:** This is a broad-based term, coined at the MIT Auto-ID Center, which connotes Internet connectivity to physical objects. Real-time location systems (RTLS), global positioning systems (GPS), and wireless sensor networks (WSNs) contribute to the Internet of Things.

**Interoperability:** The ability of systems, from different vendors, to execute bi-directional data exchange functions in a manner that allows them to operate effectively together.

**Interrogation:** The process of communicating with, and reading, a transponder.

**Interrogation zone:** This is the region in which a transponder or group of transponders can be effectively read by an associated radio frequency identification reader/interrogator.

**Interrogator:** A fixed or mobile data capture and identification device using a radio frequency electromagnetic field to stimulate and effect a modulated data response from a transponder or group of transponders present in the interrogation zone. Often used as an alternative term to reader. See also Reader.

**ISO 10536:** The international standard for proximity cards.

**ISO 11784:** The international standard defining frequencies, baud rate, bit coding, and data structures of the transponders used for animal identification.

**ISO 14443:** The set of international standards covering proximity smart cards.

**ISO 15693:** The international standard for vicinity smart cards.

**ISO 18000:** This is a series of international standards for the air interface protocol used in RFID systems for tagging goods within the supply chain.

**ISO 7816:** A set of international standards covering the basic characteristics of smart cards, such as physical and electrical characteristics, communication protocols, and others.

**Item-level:** This term is used to describe the tagging of individual products, as opposed to case-level and pallet-level tagging.

**iWIP:** Work-in-process visibility over the Internet, an intranet, or virtual private network (VPN), it usually involves contactless technology such as RFID.

# L

**Label applicator:** A device that applies labels to cases or other items, some label applicators can print bar codes on and encode RFID transponders in labels before applying the labels.

**License plate:** This term generally applies to a simple RFID that has only a serial number that is associated with information in a database. The Auto-ID Center promoted the concept as a way to simplify the tag and reduce the cost.

**Linear-polarized antenna:** An antenna that focuses the radio energy from the reader in one orientation or polarity, increases the read distance possible, and can provide greater penetration through dense materials. Tags designed to be used with a linear polarized reader antenna must be aligned with the reader antenna in order to be read.

**Low frequency:** From 30 kHz to 300 kHz, low-frequency tags typically operate at 125 kHz or 134 kHz. The main disadvantages of low-frequency tags are they have to be read from within three feet and the rate of data transfer is slow. But they are less subject to interference than UHF tags.

**Low-frequency tags:** RFID systems that operate at about 125 kHz with a typical maximum read range of up to 20 inches (508 mm).

# M

**Manufacturers tag ID (MfrTagID):** A reference number that uniquely identifies the tag.

**Memory:** A means of storing data in electronic form. A variety of random access (RAM), read-only (ROM), Write Once–Read Many (WORM), and read/write (RW) memory devices can be distinguished. In RFID terms, it's the amount of data that can be stored on the microchip in an RFID tag. It can range from 64 bits to 2 kilobytes or more on passive tags.

**Memory block:** Memory on the microchip in an RFID tag is usually divided into sections, which can be read or written to individually. Some blocks might be locked, so data can't be overwritten, whereas others are not.

**Memory modules:** A colloquial term for a read/write or reprogrammable transponder.

**Microprocessor:** The silicon chip that is the heart of a computing system, it includes a central processing unit, internal registers, control logic, and bus interfaces to external memory and input–output ports. Some advanced systems also include floating point processors and some memory.

**Microwave:** A high-frequency electromagnetic wave, a microwave is 1 mm to 1 m in wavelength.

**Microwave tags:** A term that is sometimes used to refer to RFID tags that operate at 5.8 GHz. They have very high transfer rates and can be read from as far as 30 feet away, but they use a lot of power and are expensive. (Some people refer to any tag that operates above about 415 MHz as a microwave tag.)

**Middleware:** In the RFID world, this term is generally used to refer to software that resides on a server between readers and enterprise applications. The middleware is used to filter data and pass on only useful information to enterprise applications. Some middleware can also be used to manage readers on a network.

**Misread:** A condition that exists when the data retrieved by the reader/interrogator is different from the corresponding data within the transponder.

**Modulation:** A term denoting the process of superimposing (modulating) channel encoded data or signals onto a radio frequency carrier to enable the data to be effectively coupled or propagated across an air interface, it is also used as an associative term for methods used to modulate carrier waves. Methods generally rely on the variation of key parameter values of amplitude, frequency, or phase. Digital modulation methods principally feature amplitude shift keying (ASK), frequency shift keying (FSK), phase shift keying (PSK), or variants.

**Modulation index:** The size of variation of the modulation parameter (amplitude, frequency, or phase) exhibited in the modulation waveform.

**Multiple reading:** The process or capability of a radio frequency identification reader/interrogator to read a number of transponders present within the system's interrogation zone at the same time. Alternative term for Batch reading.

**Multiplexor (Multiplexer):** This device connects a number of data communication channels and combines the separate channel signals into one composite stream for onward transmission through a single link to a central data processor or information management system. At its destination the multiplexed stream is demultiplexed to separate the constituent signals. Multiplexors are similar to concentrators in many respects, a distinction being that concentrators usually have a buffering capability to "queue" inputs that would otherwise exceed transmission capacity.

# N

**National Institute for Standards and Technology (NIST):** An American standards body that establishes standards for information-processing technology, particularly IT used by the federal government.

**Near-field communication:** RFID reader antennas emit electromagnetic radiation (radio waves). If an RFID tag is within full wavelength of the reader, it is sometimes said to be in the "near field" (as with many RFID terms, definitions are not precise). If it is more than the distance of one full wavelength away, it is said to be in the "far field." The near-field signal decays as the cube of distance from the antenna, whereas the far-field signal decays as the square of the distance from the antenna. So passive RFID systems that rely on near-field communication (typically low- and high-frequency systems) have a shorter read range than those that use far-field communication (UHF and microwave systems).

**Noise:** The unwanted extraneous electromagnetic signals encountered within the environment, usually exhibiting random or wideband characteristics, and viewed as a possible source of errors through influence upon system performance. Compare Interference.

**Noise immunity:** A measure of the extent or capability of a system to operate effectively in the presence of noise.

**Nominal range:** This is the normal range at which a system can operate reliably, under normal conditions.

## O

**Omnidirectional:** A description of a transponder's ability to be read in any orientation.

**Open systems:** Within the context of radio frequency identification, they are systems in which data handling, including capture, storage, and communication, are determined by agreed standards, thus allowing various and different users to operate without reference to a central control facility. Compare with Closed systems.

**Orientation:** The attitude of a transponder with respect to the antenna, expressed in three-dimensional angular terms, with range of variation expressed in terms of skew, pitch, and roll.

**Orientation sensitivity:** The sensitivity of response for a transponder expressed as a function of angular variation or orientation.

**Out-of-field reporting:** A mode of operation in which a transponder is identified as having left the reader interrogation zone.

# P

**Parity:** A simple error-detecting technique, it is used to detect data transmission errors, in which an extra bit (0 or 1) is added to each binary represented character to achieve an even number of 1 bits (even parity) or an odd number of 1 bits (odd parity). By checking the parity of the characters received a single error can be detected. The same principle can be applied to blocks of binary data.

**Passive transponder (tag):** A battery-free data-carrying device that reacts to a specific, reader produced, inductively coupled or radiated electromagnetic field, by delivering a data-modulated radio frequency response. Having no internal power source, passive transponders derive the power they require to respond from the reader/interrogator's electromagnetic field. Compare Active (transponder) tag.

**Penetration:** This term is used to indicate the ability of electromagnetic waves to propagate into or through materials. Nonconducting materials are essentially transparent to electromagnetic waves, but absorption mechanisms, particularly at higher frequencies, reduce the amount of energy propagating through the material. Metals constitute good reflectors for freely propagating electromagnetic waves, with very little of an incident wave being able to propagate into the metal surface.

**Phase modulation (PM):** A representation of data or signal states by the phase of a fixed frequency sinusoidal carrier wave. Where data is in binary form the modulation involves a phase difference of 180° between the binary states and is referred to as phase shift keying (PSK).

**Phase shift keying (PSK):** A representation of binary data states, 0 and 1, by the phase of a fixed frequency sinusoidal carrier wave, a difference of 180° being used to represent the respective values.

**Polar field diagram:** A graphical representation of the electric or magnetic field intensity components of an electromagnetic field, expressed on a polar co-ordinate system (distance vs. angle, through 360°), it is typically used to illustrate the field characteristics of an antenna.

**Polarization:** The locus or path described by the electric field vector of an electromagnetic wave, with respect to time.

**Port concentrator:** This is a device that accepts the outputs from a number of data communication interfaces for onward transmission into a communications network.

**Power levels:** A measure of the amount of RF energy radiated from a reader or active tag, it is usually measured in volts/meter.

**Printer/Encoder:** The device used to generate smart labels, they both print bar-coded labels and encode RFID tags embedded in the labels. See also Smart label.

**Programmability:** The ability to enter data and to change data stored in a transponder or tag.

**Programmer:** An electronic device for entering or changing (programming) data in a transponder, usually via a close proximity, inductively coupled, data transfer link.

**Programming:** The act of entering or changing data stored in a transponder or tag.

**Projected lifetime:** The estimated lifetime for a transponder often expressed in terms of read or write cycles or, for active transponders, years, based upon battery life expectancy and, as appropriate, read/write activity.

**Protocol:** A set of rules governing a particular function, such as the flow of data/information in a communication system.

**Proximity:** This term is often used to indicate closeness of one system component with respect to another, such as that of a transponder with respect to a reader.

**Proximity sensor:** An electronic device that detects and signals the presence of a selected object, and when used in association with a radio frequency identification system the sensor is set up to sense the presence of a tagged or transponder-carrying object when it enters the vicinity of the reader/interrogator so that the reader can then be activated to effect a read.

**Pulse dispersion:** The spread in width or duration of a pulse during transmission through a practical transmission system, due to the influence of distributed reactive components.

# R

**Radio frequency identification system (RFID):** An automatic identification and data capture system composed of one or more reader/interrogators and one or more transponders in which data transfer is achieved by means of suitably modulated inductive or radiating electromagnetic carriers.

**Radio frequency tag:** An alternative term for a transponder.

**Range–programming:** The maximum distance between the antenna of a reader/interrogator and a transponder over which a programming function can be effectively performed is the range–programming. It is usually shorter than the read range, but may be influenced by orientation and angle with respect to the antenna, and possibly by environmental conditions.

**Range–read:** The maximum distance between the antenna of a reader/interrogator and a transponder over which the read function can be effectively performed is the range–read. The distance will be influenced by orientation and angle with respect to the antenna, and possibly by environmental conditions.

**Read:** The process of retrieving data from a transponder and, as appropriate, the contention and error-control management, and channel and source decoding required to recover and communicate the data entered at source.

**Read only:** This term is applied to a transponder in which the data is stored in an unchangeable manner and can therefore only be read.

**Read rate:** The maximum rate at which data can be communicated between transponder and reader/interrogator, usually expressed in bits per second (bps).

**Read/write:** Applied to a radio frequency identification system, it is the ability to both read data from a transponder and to change data (write process) using a suitable programming device. See Reader/interrogator.

**Readability:** The ability to retrieve data under specified conditions is called the readability.

**Reader/interrogator or reader/writer:** An electronic device for performing the process of retrieving data from a transponder and, as appropriate, the contention and error-control management, and channel and source decoding required to recover and communicate the data entered at the source. The device may also interface with an integral display or provide a parallel or serial communications interface to a host computer or industrial controller.

**Redundancy:** In information terms it describes the additional bits, such as those for error control or repeated data, over and above those required for transmitting the information message.

**RF tag:** An alternative shorthand term for a transponder.

**RS232:** A common physical interface standard specified by the Electronics Industries Association (EIA) for the interconnection of devices, the standard allows for a single device to be connected (point-to-point) at baud values up to 9,600 bps, at distances up to 15 meters. More recent implementations of the standard may allow higher baud values and greater distances.

**RS422:** A balanced interface standard similar to RS232, but using differential voltages across twisted pair cables, it exhibits greater noise immunity than RS232 and can be used to connect single or multiple devices to a master unit, at distances up to 3,000 meters.

**RS485:** An enhanced version of RS422, it permits multiple devices (typically 32) to be attached to a two-wire bus at distances of over 1 kilometer.

## S

**Scanner:** The combination of antenna, transmitter (or exciter), and receiver into a single unit is often referred to as a scanner. With the addition of electronics to perform the necessary decoding and management functions to deliver the source data, the unit becomes a reader.

**Sensor:** An electronic device that senses a physical entity and delivers an electronic signal that can be used for control purposes.

**Separation:** The operational distance between two transponders.

**Smart label:** A label that usually contains both a traditional bar code and an RFID tag. As bar codes are printed on smart labels, information is also encoded into the RFID tag by the printer.

**Source decoding:** The process of recovering the original or source data from a received source encoded bit stream.

**Source encoding:** The process of operating upon original or source data to produce an encoded message for transmission.

**Spectrum–electromagnetic:** The continuum of electromagnetic waves, distinguished by frequency components and bands that exhibit particular features or have been used for particular applications, including radio, microwave, ultraviolet, visual, infrared, X-rays, and gamma rays.

**Spectrum–signal:** The make-up of a signal or waveform in terms of sinusoidal components of different frequency and phase relationship (spectral components).

**SRD (Short-range device):** A tag that is used at short range (less than 100 mm).

**Synchronization:** The process of controlling the transmission of data using a separate or derived clocking signal.

**Synchronous transmission:** A method of data transmission that requires timing or clocking information in addition to data.

# T

**Tag:** This colloquial term for a transponder is commonly used and is the term preferred by AIM for general usage.

**Tag antenna:** The conductive element that enables the tag to send and receive data. Passive, low- (135 kHz), and high-frequency (13.56 MHz) tags usually have a coiled antenna that couples with the coiled antenna of the reader to form a magnetic field. UHF tag antennas can have a variety of shapes. Readers also have antennas that are used to emit radio waves. The RF energy from the reader antenna is "harvested" by the antenna and used to power up the microchip, which then changes the electrical load on the antenna to reflect back its own signals.

**Time suck:** Slang for any time-consuming application (usually social networks) for no purpose other than entertainment or socializing.

**Tolerance:** The maximum permissible deviation of a system parameter value, caused by any system or environmental influence or impact. It is usually expressed in parts per million (ppm). Tolerances are specified for a number of radio frequency parameters, including carrier frequencies, subcarriers, bit clocks, and symbol clocks.

**Transceiver:** TRANSmitter/reCEIVER device used to both receive and transmit data. See also Transmitter. Compare Transponder.

**Transmitter (exciter):** An electronic device for launching an electromagnetic wave or delivering an electromagnetic field for the purpose of transmitting or communicating energy or modulated data/information, it is often considered separately from the antenna, as the means whereby the antenna is energized. In this respect it is also referred to as an exciter.

**Transponder:** An electronic TRANSmitter/resPONDER, commonly referred to as a Tag.

# U

**UltraHigh-Frequency (UHF) tags:** RFID system that operate at multiple frequencies, including 868 MHz (in Europe), a band centered at 915 MHz, and 2.45 GHz (microwave). Read range is typically 3 to 10 feet (1 to 3 meters), but systems operating in the 915 MHz band may achieve read ranges of 20 feet (6 meters) or more.

**Uniform Code Council (UCC):** This organization in the United States sets and maintains the Universal Product Code (UPC) bar-code standard.

**Universal Product Code (UPC):** The bar-code standard used in North America. See also Uniform Code Council.

**Uplink:** This term defines the direction of communications as being from transponder to reader/interrogator.

# V

**Vector:** A quantitative component that exhibits magnitude, direction, and sense.

**Verification:** The process of ensuring that an intended operation has been performed.

# W

**Write:** The process of transferring data to a transponder, the internal actions of storing the data, which may also encompass the reading of data to verify the data content.

**Write Once Read Many (WORM):** This distinguishes a transponder that can be part or totally programmed once by the user, and thereafter only read.

**Write rate:** The rate at which data is transferred to a transponder and stored within the memory of the device and verified. The rate is usually expressed as the average number of bits or bytes per second over which the complete transfer is performed.

## X

**XML:** eXtensible Markup Language, a widely accepted way of sharing information over the Internet in a standard generic way.

# Appendix C: Standards Used in Wireless Enterprises

One challenge facing companies deploying wireless sensors is the disparate standards, protocols, and methods of communication and data formats.

Wireless network standards organizations such as ZigBee, WirelessHART, and the ISA SP100 are all working to make wireless standards and products more appropriate for industrial applications. But standards bodies are typically slow to move and it is unlikely in the long term that any one company will use a single standard for all applications.

Standards are very important for the adoption of wireless sensor technology: creation of standards typically leads to an explosion of end products built to that standard which is a necessity for any emerging market (e.g., of RFID tags to the EPCglobal Class 1 Generation 2 standards). However, the confusion around competing standards and which is the best for a given application typically leads to inaction and market roadblocks. Interoperability across any and all standards will help drive the adoption of wireless sensors.

A second challenge is that there is currently no data format standard. The methods of data packet transfer are defined in these standards; however, the method and format of data sent out of the network are not. On a ZigBee network, the transmission rate, power consumption, and embedded stack configuration are all included in the standard, but how the data is actually delivered into an application is left up to the developing company. This can add confusion for a company that is deploying ZigBee products from two vendors, as the data structure may not be the same.

## TABLE C.1

Short List of Standards Used in Wireless Sensor Networking

| Standard | Description |
|---|---|
| 802.11 | A family of specifications developed by IEEE for local area networks. Typically high bandwidth and high-speed data rates and larger data packets. |
| 802.15 | A family of specifications developed by IEEE for personal-area networks. Typically low power, low rates, and small data packet size. |
| Bluetooth | A short-range wireless standard operating in the unlicensed 2.4 GHz spectrum. |
| ISA SP100 | An open industrial wireless standard trying to support multiple protocols in a single standard. Includes 2.4 GHz and 802.14.5 radios. |
| Wi-Fi | Wireless Fidelity: technologies based on the 802.11 standard. |
| WiMax | World Interoperability for Microwave Access: a standard for wireless broadband over long distances based on IEEE 802.16. |
| WirelessHART | An open wireless communication standard from the HART Communications Foundation designed for process measurement and control applications. Based on the 802.15 standard and frequency hopping spread spectrum technology. |
| ZigBee | A low-data-rate, two-way standard for home automation and data networks. Uses very low power consumption to create mesh networks using 802.14 radios. |
| Zwave | Low-power, low-bandwidth communications standard designed for interoperability between systems and devices. Geared toward the residential and light commercial devices. |

# Appendix D:
# Lean Wireless ROI

Technology makes it possible for people to gain control over everything, except over technology.

**John Tudor**

Just 'cuz I don't understand you don't mean you're brilliant.

**Cowboy aphorism**

No responsible CEO or operations manager will allow a Lean Wireless transformation without being certain that (1) it pays for itself, and (2) it does not disrupt business as usual; rather, it allows more business than usual, including higher productivity, faster patient diagnosis, zero stock-outs, or fewer wrong turns on the highway by truckers. Every gain in efficiency saves something, be it direct costs; the indirect costs of labor, time, and space; or the really indirect costs of replacing employees or marketing to replace lost business. And yet that CEO or operations manager must be convinced of ROI, else an enterprise risks being left behind on the technology curve.

In this Appendix, and in the Lean Wireless ROI Calculator that accompanies the book (at www.leanwireless.com), we show just how Lean Wireless returns money to your organization. Your purchase of *Thin Air* entitles you to use this one-of-a-kind resource. The Calculator is easy to use, but uses the complex calculations and methods outlined in this chapter; think of this as a "Kiplinger TaxCut"® of Wireless. Please allow a half hour of your time, visit www.leanwireless.com, and calculate the savings that Wireless represents to your company.

Of course, the Lean Wireless ROI Calculator is not as exact as the thorough, expert analysis we describe in the following pages. Calculating ROI is meticulous, pedantic, and sometimes picayune. But, the rewards (typically 250-500 percent ROI) are worth it. We attempt in the following pages to make that process as clear as possible.

## HOW PRECISELY CAN WE MEASURE THE ROI OF LEAN WIRELESS?

Truth be told, we can't. In Chapter 1, "Lean Wireless Is Already Here," we compared the step changes of Wireless as being historically equivalent to the telephone, radio, jet flight, plastics, and the like. (Millennial workers like to refer to such advances as "game changing.")

What is the ROI of your telephone system; or of your local area network? Does any leader at your organization need to be convinced that there is one, in order to keep it? It is similarly difficult (and purposeless) to calculate the ROI of Wireless Communication on an organization, taking into account every employee's use of Wireless Communication technology, including cellular phones, PDAs, notebook computers, and Wi-Fi.

What those leaders are likely less convinced of is the ROI of Tactical Wireless, such as RFID, RTLS, WSNs, and GPS. As with the ROI on Lean practices, some elements of Wireless ROI are perfectly measurable—you can put a dollar sign in front of them—but not all. Or, their fiscal returns take time to reveal themselves.

For example, two objectives of continuous improvement are customer satisfaction and worker satisfaction. These seem like soft-edged goals, but the ROI is found in retention of both employees and customers, which is not "funny money." The authors of *E-Marketing* (now in its fifth edition) estimate that it costs five times as much to gain a customer than to retain one.[1] An oft-cited Berkeley University study on the effects of the U.S. Family Medical Leave Act estimates that turnover costs for a manager average 150 percent of salary, between both the tangible costs (hiring new workers, relocation) and such intangibles as lost efficiency and profit while in the interim.[2]

A good many returns on a Tactical Wireless investment are wonderfully measurable, and quick. Recall that fully 75 percent of companies using RFID expect a return on investment within 18 months, according to an ABI research study.[3] They have every reason to expect that. As we saw in Chapter 4, Mercy Medical's Catheterization Laboratory saw a 500 percent ROI in 18 months. That return included reducing inventory levels by $376,587 (25 percent) in the first half of 2008, and reducing waste from expired products by 40 percent. Mercy Medical also saved 1.5 hours per weekday, or 7.5 hours per week, in conducting manual inventory.

The ABI study found that companies that have implemented RFID into production systems observed such measureable benefits in the areas of:[4]

- Labor reduction
- Inventory visibility
- Efficiency gains/business process improvement
- Asset visibility
- Physical security
- Achieving regulatory compliance
- Achieving customer compliance
- Logical security
- Advanced/automatic decision support

That is an excellent short list of KPIs (key performance indicators) to measure.

## THE BARRIER OF DISTRUST

There are dozens of clever methodologies out there from true industry wizards, and we have collected some of them here and in the accompanying "ROI Calculator for RFID, RTLS, and Wireless Sensors." The ROI Calculator cannot take into account every single dollar of return—perhaps not even most—and this, we believe, is the single greatest barrier to companies that implement Wireless, and gets in the way of such Wireless technologies as RFID. In the 2008 ABI Research study, nearly 61 percent of respondents did not use RFID.[5] The primary reason they gave was they were already using some other form of auto-ID (such as bar coding). But other important factors boiled down to a lack of demonstrable ROI, a lack of understanding of the benefits, and the cost of equipment and services. But if the tangible ROI exceeds 100 percent in six months, an exact figure hardly matters.

Wireless technology improves three elements in an enterprise: (1) visibility, (2) control, and (3) communication. Applied properly, Wireless provides measurable benefits that offset the total cost of ownership, or TCO. Enterprises at large may harbor some distrust of the technology, but entities such as Mercy Medical, the Stobie Mine, American Airlines, and SAP have all calculated the ROI of Wireless, and found it to exceed 100 percent.

## THE COMPONENTS OF ANY COST

The phrase "value stream" connotes a linear process. It's a useful way to think of it, but really, a value stream is a series of interrelated activities: if you change one activity, all other activities on the value stream (and several off the stream) are affected. For example, if a mechanic fails to secure a needed engine part for replacement, then every value stream that follows is delayed, including returning the car to the customer. So also are two key processes off the value stream, being charge capture and billing.

If a given activity seems difficult to monetize (such as making a decision, affixing a signature, searching for a hardcopy file), consider that any activity includes four perfectly measurable elements, and that each of these elements represents cost. Therefore, an effective ROI model converts these four components into cost. These four components are as follows:

- *Time.* Only in time do change, value addition, delivery, and profit occur. Time can be converted to cash via unit cost and hourly wage. If a $50/hour employee spends five minutes making a decision, that decision cost the organization $4.16.
- *People*, such as that employee, who are the agents of change. Only through the direct intervention of, for example, a mechanic, doctor, customer service representative, machine operator, or forklift driver is time used effectively to create value.
- *Objects*, which are the creations of people. Objects may be used to execute a process (objects such as tools, forklifts, computers, or scalpels), or may be the product of a process (a finished good on a store shelf or an automobile in a dealer's lot).
- *Space* wherein either an activity occurs, or the product of activities occurs. In terms of activity, space can be a factory floor, cubicle, car dealership, restaurant, retail store, shipping route, and so on; in terms of product of activity, space is the area consumed by inventory on a factory floor or in a distribution center, on a store shelf, and the like.

In a perfect world, every one of those components would expand and contract in response to demand, which, to a degree, they can. Retailers expand and contract Component 2—People—by using seasonal help. General contractors expand or contract Component 3—Objects—by renting equipment such as backhoes that they use infrequently.

Compressing Components 1 and 4—Time and Space—seems absurd at first glance, but far from it. Lean and Wireless both aim to shorten time to fulfillment along a given value stream, be it delivering a finished automobile from an assembly line, or treating a patient in an emergency room. Both Lean and Wireless also attempt to compress space, not just by freeing space required by inventory, but, in allowing for the reconfiguration of a workspace for greater efficiency. Compress time and space, and you have freed capacity for more work. Thus Component 2—People—is capable of a greater yield, and higher productivity.

The simplest ways to measure the four components are these:

- The cost of inventory can be measured by cost of goods sold (COGS).
- Equipment assets must include, at minimum, depreciation and base cost. It would also include maintenance, repair, and operating supplies (MRO), costs that include all supplies (disposable and nondisposable) used in a manufacturing process.
- Space is measured in square footage or square meters.
- Time and people measurements are based upon an average hourly wage of the labor type.

Every business process consumes one or more of the four elements of cost. Similarly, every business process has three stages: (1) an input, (2) one or more activities, and (3) an output. Billing is a business process, involving (1) confirmation of a shipment; (2) figuring payment due, taxes, tariffs, and the like; and (3) generating an invoice. Because Stage 2, Activities, can be expressed as one of the above four components (time, people, objects, and space), and those components can be expressed as costs, then every business process—no matter how abstract—can be expressed in terms of costs.

The most simple, straightforward, and generic processes are normally the best place to find ROI.

---

## PRESUME VALUE (FOR THE SAKE OF ARGUMENT), AND FORGE AHEAD

Earlier, we asked if any executive at your enterprise needed to be convinced of the utility of your phone system, or your LAN. There is no question whatsoever that either is required, and no one has conducted an ROI analysis of either. But as an exercise, consider how your employees would communicate internally without either; through interdepartmental mail, with dedicated employees moving mail carts? Leaving their workstations to speak to others, within reasonable walking distance? Letters and telegraphs between branches? How much would the added labor cost, and, how much business would it cost the company, now that it is stripped of the most modern tools?

If rather than asking what Wireless is and if it has value, you ask what Wireless does and presume value, then it becomes an industrial engineering (IE) or cost engineering (CE) question, with a fairly well-defined method of approach.

Cost engineering principles look at the process from a standpoint of efficiency gains, such as touch point reduction, process transformation, and productivity. It also means looking at the human and asset aspects from an accounting standpoint, as both direct and indirect costs. Finally, the approach considers qualitative cost or "soft costs" that occur as a result of employing Wireless.

Put all the costs together and you have "Total Cost Management" (TCM) or because we are talking about equipment, "Total Cost of Ownership" (TCO). The technology must produce positive bottom-line results in excess of TCM or TCO.

The IE/CE approach entails three categories: (1) process evaluation, (2) technical evaluation, and (3) financial evaluation. These can be applied to every process and subprocess that you want to evaluate for Wireless.

A process evaluation must consider more than just the Wireless-enabled process. It must consider the downstream, upstream, and parallel processes as well, leaving no stone unturned. It has to look at the ergonomics of activities as well as the methods. For example:

- Where does manual data capture occur?
- How are exceptions handled and how frequently do they occur? How is data used? What technology platform is used?
- What does the data drive?
- Is this a manual process or an automated process?
- What are the step-by-step functions of the operator setup process?
- Where are the visibility gaps in the process?

- What are the process constraints?
- How is inventory used or affected by these constraints?

The questions then must escalate to the larger questions:

- What are the strategies of the company?
- Is this part of a greater technological transformation?
- What would the impact of a new architecture be on the system, or will this be an integrated architecture?
- How much integration and customization is necessary?
- How will this affect our customer relationships and contracts?
- What are the inventory and purchasing strategies?
- Is there lead-time mitigation?
- Is there an expansion strategy?

The second step is to determine what, if any, technology platform can support the process changes identified during the process feasibility phase. Jumping on the RFID bandwagon for the sake of having the latest technology does a disservice to shareholder value. Instead, take a more thorough, painstaking, and professional evaluation of the available technologies to see if a good solid fit can be achieved.

Finally, find the money. The most effective way to do a financial evaluation is to either measure by pain or measure by value, which are not the same, and not necessarily related. If a pain is document tracking, and the value is shipping management, it might be best to evaluate the pain for the pilot and the value for the longer-term solution (more on that later).

## DEFINE YOUR BUSINESS WITH BASELINING

Continuous improvement calls for an honest evaluation of the existing, or "as-is," process by the people involved in that process. That evaluation is used as a baseline for improvement. Baselining in Lean Wireless begins just as it does in Lean, by gathering together the stakeholders, key people from the areas that are directly affected. In a supply chain evaluation, this would include knowledgeable representatives of shipping and receiving, supply chain management, operations, information technology (IT), procurement, sales and marketing, customer service, and a comptroller or accounting manager. This must also include a project manager, whose skills will be required in managing this complex process.

The usual baseline procedure is as follows.

## Step 1: Conduct a Whiteboard Study

Start on a whiteboard or overhead projector. List all the areas that are affected by or being considered for Wireless.

Add the processes that directly interlace with what is listed. This is vital to understanding the parallel, upstream, and downstream processes affected by the processes that will change.

For each process listed, have the person responsible for that process map out each step. Obviously, this will be time consuming, but this is how the impact and ROI of any improvement is measured.

## Step 2: Conduct a DILO Study

After high-level process maps are complete, conduct a "reality check" by conducting a day-in-the-life-of study (DILO). A DILO is an informal time and ergonomic study that observes a typical day, measures how long it takes to perform tasks (people typically underestimate by half), counts the touch points, and reveals how breaks and changeovers affect productivity.

The importance of the DILO step cannot be emphasized enough. All the brainstorming in the world cannot replace walking unannounced through the process on an ordinary workday. The objective is not to find blame for process faults, or even to identify the faults; rather, the objective is to define what exactly the existing process is. A DILO should involve middle managers and supervisors to understand the whys and what is involved in any process.

A DILO is not the same as a full-time study; this is a Six Sigma-like method involving full-time observation (manual or automatic) and thousands of measurements over a length of time. It delivers more accurate results than a DILO study, but is probably overkill; most companies do not have the budget or time to conduct a full-time study, besides which, a DILO is sufficient, allowing you to extrapolate savings as a percentage.

A DILO study must capture the four parameters with which we opened the chapter: time, people, objects, and space.

In observing time, ask what tasks are performed and how long does each take. What is the throughput rate, the cycle time for the segment, and the takt time average? (Takt time is the maximum time allowed to produce a product in order to meet demand.) Look for natural time breaks and unnatural time breaks. How much time is spent recording information, verifying information, or even in communication?

In observing people, ask how many people are on shift and perform the task. How many touch points are there? How frequently is a person pulled off task to perform supporting functions such as searches or data entry? What is the turnover rate? Be sure to ask individuals what they do and how they do it. Tell them that you are looking for ways to make their work easier and more productive, and ask them about their "pain." Understand though how the culture works in your organization and how the direct interaction should be handled. (A challenge that continuous improvement consultants have always faced is suspicion; will efficiency measures eliminate jobs? The best answer is that it will free up capacity for more work.)

In quantifying objects, observe what equipment is used in a process. Be acutely aware of the objects of labor. Do operators in a warehouse rely upon touch screens or bar-code scanners, or both? Objects have a way of being far costlier than they appear; a typical fat-client desktop computer costs about $1,025 per client per year, between acquisition, deployment, and maintenance.[6] This is startling, considering that the equipment alone costs little more than $1,000. Thus a $1,000 installation in use for four years costs $4,200. Similarly, the TCO of a forklift involves fuel and operations costs, as well as maintenance and depreciation costs.

Finally, in observing space, chart and measure the ergonomic flow of a process. For purposes of the DILO study, limit yourself to physical space, versus cyberspace. WIP on a production floor can be measured in feet, but WIP in an accounting department is almost entirely electronic. This is a good time to obtain the industrial drawings or architectural drawings of the physical areas wherein the processes occur. Also consider the ergonomic environment such as object placement, available space for travel during busy times, and so on. Be certain to mark exception, departure, and arrival areas for all processes. This information will be required in the next stage: the technical analysis.

## Step 3: Reconcile Process Maps and DILO

Combining the findings from the DILO with the high-level process maps from the whiteboard session creates a realistic analysis of baseline processes. Some common forms are these:

- Process maps (traditional flowchart)
- Multilayer process maps (with multiple interlaced processes depicted by color)
- SIPOC (supplier input process output customer)
- Spaghetti diagrams (which is the actual movement process you observed in the DILO)
- Process value maps (with data listed on the process map)

## Step 4: Target a Short List of Process Improvements

Recall that Lean Wireless adds value when it eliminates or compresses such cost factors as time, unnecessary manual work, equipment, surplus inventories, or space used that could readily be converted to value-added capacity.

Of all of the processes mapped in Steps 1 through 3, select perhaps three processes with the greatest pain, then three strong opportunities for Lean Wireless value, ones that will benefit by improved visibility or connectivity.

The most common pain found in any organization is what is typically referred to as an exception. An exception is something that causes deviation from the normal process. Lean Wireless technologies are very often used to reduce exceptions and improve exception handling. Very commonly, exceptions are caused as a result of error during a routine manual process, such as forgetting to scan a bar code or scanning the same bar code twice. The resulting error will have a cost (or pain) associated with it. For example, not counting the item or counting it twice. It is important to calculate the average cost of an exception for a given process, as well as the cost for handling that exception. (Likely, your team identified these in the whiteboard analysis in Step 1.)

## Step 5: Choose between Proof of Concept or Full Steam Ahead

Once you have targeted those six opportunities, tentatively decide if your goal is to move on to a proof of concept, or simply to implement Wireless technology.

This may vary by technology. You likely do not need to be sold on the advantages of GPS in company vehicles, or remote and wireless sensors on key equipment, but RFID and RTLS, or elaborate wireless sensor networks, may be another matter.

If you choose proof of concept, consider focusing upon eliminating pain points; many RFID integrators have learned that it is quicker and less complex to solve inefficiencies than to create value. But it is important to recognize that the proof of concept tells only half the story; pain elimination will need to cover costs and fall within your organization's modified internal rate of return (MIRR) requirements. (MIRR is a financial measure used to determine the attractiveness of an investment. It is generally used as part of a capital budgeting process to rank various alternative choices.)

If your organization is looking at RFID from an operational use standpoint, then you should still conduct a proof of concept; however, the driving factor behind it should be creating value. The pilot should maximize value to acquire the strongest return gains in the shortest possible time frame.

## CONDUCT A TECHNICAL EVALUATION

Recall that the value propositions of Lean Wireless are to increase visibility and to improve efficiency, control, and execution of your processes; these capabilities in turn should enable revenue generation and cost management. Any technology that does not fulfill those criteria has no ROI.

A technical evaluation determines what, if any, technology platform can fulfill those criteria. The details of the technical decisions have significant impact on future capital decisions when considered from a cost and value perspective. (See sidebar, "How Not to Install RFID: 15 Blunders.") We encouraged a bit of faith, earlier in the chapter, but also advised that adopting the technology for the sake of having the latest technology can cost more than it returns. Instead, take a more thorough, painstaking, and professional evaluation of the available technologies to see if a good solid fit can be achieved.

### Step 1: Conduct an Environmental Analysis

The enterprise environment is far more than its physical location; rather, it is its entire "ecosystem," within its four walls, over wires, in the market, and across its supply chain. Wireless can create value across the entire ecosystem (or, poorly implemented, create havoc just as easily).

A complete environmental analysis examines five elements:

- Physical
- Ergonomic
- IT and infrastructure
- Customer
- Competitive/strategic

Each of these five areas has a direct impact upon the ROI of any solution, and figures in determining which Wireless solution makes the most sense.

The physical environment is where you intend to deploy the technology. In the case of RFID, this is not limited to manufacturing lines, dock doors, or conveyors, but includes any products and assets to be tagged. Equipment, tags, packaging, and power sources make up this portion of the analysis. Cost is found in potential changes such as packaging, branding marks, drilling or cutting, replacing of rollers, installation of power drops, and so on.

The ergonomic environment is the point at which the process information you gathered previously in the DILO is inserted into consideration. OSHA, work space, clearance, and work centers must be considered. The cost here again is change: change in space, change in object placement, change in methods and setup processes, and in people's work tasks. How a solution is designed to fit ergonomically, and the change it imposes upon your labor force, will directly affect productivity.

The IT environment and infrastructure have a key role within the analysis. RFID and RTLS systems are visibility sensor networks that your system leverages for near-real-time data. Therefore, the system falls squarely between the Operations and Information Technology groups. Like any good information system, the system must be aligned squarely with a business's best practices and strategy. Table D.1 lists numerous considerations and cost factors to consider.

Your end customer must also be a factor. Consider contractual compliance and relationship direction as soft costs, but the loss of potential business or sales as hard costs. Compliance also ties into potential new channel development costs; however, these would not normally be expressed in a capital budgeting analysis.

**TABLE D.1**

Cost Factors and Considerations for Introducing Wireless into an IT Environment

---

**IT System Support:** What are the mechanisms and guiding operational procedures this system will fall under, and how much will it cost as it pertains to network security, information centralization, scalability, maintenance, and integration?

**IT Customer Support:** Once the Lean Wireless system is part of your normal operations, is it likely you will need specialized call center personnel or techs who are trained to handle the system? Even the most plug-and-play Lean Wireless solutions, such as PDAs, will require some IT support. This cost must be considered within the IT budget.

**Licensing:** What is the cash outlay for ongoing application and licensure? These fees will be recurring and contractually tied. For an RTLS, this would include software-as-a-service contracts; for Wireless Communication, this would include data plans.

**Forward/Retro-Compatibility:** Will the technology integrate cleanly into legacy systems? Can, for example, the RFID software system handle an upgrade in your ERP or other interfacing systems? Installing a system purely on price, only to be faced with a future expense to upgrade, can eliminate any savings.

**Platform Diversity:** If your organization has a universal IT platform, then the answer is self-explanatory; however, if you have multiple platforms, will an RFID, RTLS, or WSN system be platform-independent? Can the architecture handle Windows, Linux, and Solaris equally as well? The more platform-independent the Wireless system is, the stronger your future support position will be.

**ERP or WMS Upgrades:** If your organization is planning an ERP or WMS upgrade, you must decide to either integrate into the legacy system or buy a forward-compatible architecture that can handle the Wireless functionality in generic ERP systems.

**EDI/ASN/External Integrations:** Is your organization using electronic data interchange (EDI), some collaborative planning and forecasting system (CPFR), or similar customer/vendor-facing applications? Do you need to have advance shipping notice (ASN) transmission to conduct business, and how much will that cost?

**Lead Time (Resource, Installation):** How long will implementation take, and how much will it cost, between hiring an integrator and using internal resources?

**Standardization:** Finally, can all of this be standardized? If not, how much will the transition and extra intrasystem translation cost? The same rule applies to Wireless technology that applies to ERP; the more out-of-the-box solution is less costly over time.

---

Competitive and strategic environment should also be part of your environmental analysis, especially if you intend to leverage Wireless visibility for competitive or strategic advantage. Will Wireless connectivity give your firm a competitive, technological, or efficiency advantage? Even more important, how will your shareholders perceive the investment? In all points the deployment must be aligned with your corporate strategic goals to add sustained value.

## Step 2: Select a Suitable Technology

After the environmental evaluation is complete, it is time to evaluate the technology. For Wireless Communication, this should be fairly simple; for Tactical Wireless (including RFID, RTLS, and WSNs), evaluation must take into account performance, installation, and integration.

Tactical Wireless, as we saw in early chapters, is increasingly plug-and-play; still, evaluation of any Tactical Wireless implementation should be performed by an experienced integrator who knows the idiosyncrasies of the technology, and offers well-defined and established industry relationships and stellar client references. Be sure to work with a company that understands your specific business and has successfully implemented Lean Wireless technologies at similar companies, perhaps even your competition. (See sidebar, "How to Select a Wireless Integrator.")

The integrator will help you understand the costs of the technology, and your own audit will reveal any hidden enterprise or ongoing organizational costs, and should offer a strategy to teach your people how to maintain and expand the Tactical Wireless system, and scale it up.

As tempting as it is to do without, such an individual will know through experience:

- The technology that best suits your current budget
- The technology that suits a given commodity (e.g., high-frequency RFID where there is a high moisture content)
- Which systems best suit an enterprise's existing infrastructure (e.g., its SAP backbone)
- Which devices, such as RFID readers and wireless sensors, are truly plug-and-play

Regardless of marketing statements, the technology will work only as well as it is designed, and no better; and will work only within the limits of your unique environment. Because product designs and enterprise environments vary, there can be a dramatic difference between one system and another in the effort it takes to make the system work. More infrastructure or peripherals may be required to make the system work correctly, which adds immediate cost in equipment and ongoing costs in maintenance. A good integrator will conduct a full site survey that includes a 48-hour spectrum analysis at the facility to be installed.

**TABLE D.2**

Questions for Tactical Wireless Vendors and Integrators

| Question |
| --- |
| What levels of customization for hardware and software will be required? |
| Are there legacy integration capabilities and for what cost if required? |
| What levels of firmware upgrades and life-cycle support exist? |
| What are some of the licensing and recurring costs? |
| Are there maintenance provisions in service contracts? |
| What are past issues and performance based upon previous engagements? |
| What are the physical integration design and cost? |
| What level of scalability of the design and system exists? |
| What types of process and system change will be required and will training and support be provided? |
| How will the system handle changes in technology maturity? Will the system put in today be obsolete in two years? |

Table D.2 lists a short (but not exhaustive) list of questions for any hardware vendor or integrator; the answers will dramatically affect your TCO and ROI.

# CALCULATE THEORETICAL ROI

Thus far, the process evaluation has answered the question, "What is being done?" within your organization. The technical evaluation answered the question, "What can be done to improve business in real, tangible terms?" Now is the time to put a pencil to it, converting those abstract observations and ideas into a set of valid financial measures.

To do this properly, ROI cannot be considered from a pure accounting approach. There must be a pragmatic way to account for the broader financial costs and savings of the project. From a financial standpoint, when conducting a Benefit-Cost-Risk analysis, ask two key questions:

1. Does the ROI provide the required yield relative to the internal rate of return (IRR) for my organization?
2. Does the ROI provide the required yield relative to the risk undertaken by the project?

No matter how high the benefit, if the risk is unnecessarily high (such as a complete shutdown of manufacturing), then questions will arise as to whether the project was a good investment or just a lucky gamble.

## Step 1: Quantify the Benefits

Consider that benefits are expressed as both hard and soft categories. *Hard benefits* represent true cash inflow and true cash savings. Hard benefits will be seen directly on company financials (income, balance, and cash-flow sheets). *Soft benefits* are those that are not tied directly to cash flow, but which may alter cash flow (i.e., monies that could have been spent but were not). Soft benefits typically affect the operations budget, thus the complete ROI is not found in one place, but two.

What are the hard benefits? Let us return to the four elements of cost: time, people (labor), objects (assets), and space. Here are a few methods of modeling these elements to define the benefits.

Time usage is directly associated with all forms of productivity and throughput. It is absolutely critical to take a holistic view of a transformed system in its to-be state. Task time or cycle time improvement benefits may be translated in one or more ways:

- Increased output which may be expressed as reducing cost of goods sold (COGS)
- Equipment utilization (expressed in units of production depreciation savings)
- Touch-point time savings per unit cost (expressed in activity-based costing per unit contributions, or by per unit of labor hourly wage)

People (or labor) are similar to time savings, so be cautious not to double-count. Generally, labor savings can be expressed by factors such as:

- Time-on-task increases (time multiplied by labor hourly wage)
- Labor savings by nonexpansion of personnel (which is equal to the fully allocated yearly wage × a time period of 3 to 5 years)
- Decrease of non-value-added tasks (which is expressed as a margin percentage of COGS per unit)

Two very important considerations are retention of personnel for all businesses and the elimination of temporary personnel for companies that increase staff during peak seasons. The sum cost of a temporary employee is expressed as:

$$(\text{contractual rates} \times \text{time length of employ}) + \text{contractual riders} + \text{administrative costs}$$

Higher productivity from existing human resources should mean a company does not need as many temporary personnel.

In quantifying inventory benefits, let us concern ourselves first with inventory of the types we identified earlier, being raw materials, work in process (WIP), and finished goods.

In the case of raw materials, the objective is to measure the average holding time of raw materials, and then reduce that time. This subsequently reduces the holding costs of insurance, taxes, storage (particularly if using third-party storage), and so on. Ideally, raw materials go off the supplier's truck and directly onto the production line without any holding time. If anyone has achieved zero holding time, we have yet to hear of it. But that time can be reduced; doing so decreases the unit cost and cost of goods sold.

A measure of work in process or WIP inventory reduction is this:

$$\text{COGS} - \text{downstream transportation costs} - \text{distribution costs} - \text{holding costs} = \text{WIP inventory cost reduction}$$

WIP inventory affects the dependent MRO inventories of components. As your throughput increases, your WIP and dependent MRO inventory will both decrease. Unless your plant scraps units, reworks saved (also known as exceptions) can translate into huge savings if they can be transformed into finished inventory. Using RFID to reduce the number of reworks yielded millions of dollars a year in savings for one auto manufacturer and provided better visibility of what units were being reworked and thereby saving time to locate units. (See sidebar, "WIP Management, RFID, and ROI.")

The savings in finished goods inventory is expressed as capital savings in the form of reduced safety stock and reduced holding costs. Velocity increase would be expressed as the difference between previous velocity and new velocity multiplied by COGS.

Space measurement is tied directly to (1) capacity utilization and (2) performance. This may be expressed in two ways.

The first is the interest recovered over a time period by avoiding capital expansion, that is, getting more throughput from the current space by not having to buy, rent, or lease more space. Use it only if another capital project is planned and a Wireless implementation positively delays the need for investment due to efficiency gains. Remember, this is not the capital itself, only the interest of holding on to the funds without expenditure for a variable length of time. Again, this only applies to a committed alternative capital project affected by Wireless enhancements.

The second way is to measure the improvement of space utilization by cost/square foot/year, or by the difference in capacity performance between similar facilities, one using Wireless technology, the other not.

Soft benefits (those not directly related to cash flow) can also be defined in this analysis; however, there is no guarantee that management will be interested. They are indirectly related to cash flow, but they have cost benefits in lowering COGS and lowering operating costs; soft benefits involve the savings of increased sales relative to those costs.

They are also called "soft" because there is no clear proof that a Wireless technology such as RFID is responsible for all of a sales increase; for example, American Apparel uses RFID to ensure that goods are consistently available on its shop floor, but they still employ eye-catching well-targeted marketing campaigns. A good measure of RFID's contribution to those sales is the portion of stock-outs eliminated as a percentage of sales (multiplied by COGS).

Other soft benefits include brand equity, inventory accuracy, and the increased accuracy of forecasts (e.g., in retail, or make-to-stock manufacturing). The smaller the absolute error from the actual forecast, the more efficient production and operations will be at cutting per-unit costs by:

- Producing fewer units that do not sell (reducing buffer inventory)
- Buying only the raw material that will sell
- Reducing idle time

As we pointed out earlier, the really soft benefits that Lean and Lean Wireless foster—customer satisfaction and employee satisfaction—manifest themselves in retention.

## Step 2: Quantify the Costs

Now that you have defined the benefits, you must measure the costs. Cost comes in two flavors: conceptual and deterministic. Conceptual costs are usually done as an early "rough estimate," which you can determine using a couple of methods. The easy method includes estimating based on the historical costs of expansion.

The more complex method is a *parametric method* involving data aggregation utilizing various parameters, constraints, correlation analysis, and various forms of regression analysis. The curve that best fits the data with a strong confidence interval (R2) is generally the curve used for the estimate. Parametric data analysis is not easy, but if done correctly it will give you a strong estimate as to what costs and unknowns to expect. It is important to note that estimation accuracy will change as the project matures.

The *deterministic costing* method involves line-by-line cost estimates as close as possible, down to the nuts and bolts of the system. Within this are four subcategories: direct fixed, direct variable, indirect fixed, and indirect variable. Each must be accounted for. Examples are below:

- *Direct fixed:* Hardware, such as RFID readers, and other costs that are directly part of the projects such as applicators, antennas, and peripheral sensors.
- *Direct variable:* Tags and media. Cost is dependent upon how much you use; however, the project requires that they exist.
- *Indirect fixed:* This may be conduit, power, or network drops. They can be used for other functions; however, they are required to support the project and they do not fluctuate in cost by use.
- *Indirect variable:* Training and support provided by staff, such as maintenance or IT training. Although needed to help the project succeed, these costs do shift.

When looking at overall costs, it is best to separate out the Wireless costs from organizational costs. There are various reasons for doing this such as varying depreciation methods, capital budgeting methods, and functional cost management. Complete cost line-by-line lists are actually quite extensive. Due to the uniqueness of each deployment and environment, they are not included in this appendix. An experienced integrator working with your Wireless team and financial personnel should be able to define all of these costs.

## Step 3: Calculate the Risks (in Dollars)

Various risk methodologies exist such as failure modes and effects analysis (FMEA), multivoting, algorithmic, heuristic methods, and so on. Regardless of the method, four risk components are necessary:

- *Severity estimated value:* How severe is the problem, on a percentage scale of 1–100 percent?
- *Identification estimated value:* How easily can it be identified, on a percentage scale of 1–100 percent?
- *Mitigation estimated value:* How easily can it be fixed, on a percentage scale of 1–100 percent?
- *Cost:* How much would such a risk cost the organization in monetary terms?

The estimated values are assigned by the company and can vary over time. The formula is very simple:

$$((\% \text{ severity} \times \% \text{ identification}) / \% \text{ mitigation}) \times \$ \text{ cost} = \$ \text{ Risk}$$

Now perform this calculation for each risk line, and you have the cost of your total risk exposure.

Remember that in risk analysis, a risk such as customer loss would need to be added at full value to benefit as a savings. Evaluate risks to ensure that those representing benefit opportunities are counted on both the benefit and risk sides of the analysis.

## Step 4: At Last—Calculate ROI

After all of this, what is your ROI?

Recall these questions: (1) does the ROI provide the required yield relative to the internal rate of return (IRR) and (2) does the ROI provide the required yield relative to the risk undertaken by the project. If it fails either test, then another opportunity needs to be defined for investment, RFID or otherwise. Therefore, for an investment in Wireless to be viable:

((Benefit − Total Costs) / Total Costs) must be ≥ IRR or MIRR

((Benefit − Total Risk Exposure) / Total Risk Exposure) ≥ 3:1

ROI does exist for Lean wireless technologies, as it does for a network or IT investment, because it is much broader in reach and impact simply by improving a few processes.

You must understand your entire enterprise ecosystem (internally, electronic, and in your extended supply chain), and the transformative impact of Wireless, before trying to quantify a project.

Lean Wireless is no magic bullet, but it will bring value to most enterprise ecosystems. Trust and diligence have paid off, and handsomely, those companies that take the time and consider the process, technical, and financial components of an ROI analysis.

---

## SIDEBAR 1: HOW NOT TO INSTALL RFID: 15 BLUNDERS

Anyone can claim RFID expertise, but do they possess it? Here are some of the lessons learned the hard way by disappointed customers:

1. The integrator assumes that because she is an industry expert in a given area, she knows how to implement an RFID system.
2. The integrator orders tags that are pre-encoded but with no human readable sight verification. It is very difficult to ensure that the asset identifier (serial, asset tag, etc.) assigned to an RFID tag is assigned to the right tag and not one close to it, without this convenient second line of verification.
3. The integrator orders tags and mounts (tens of thousands of them) without first verifying that they are going to work, and that the selected method of mounting is approved by the customer. The integrator must test that mount thoroughly in as many adverse conditions as possible to ensure it will stay put. This includes the ability of a person to grab the mount and tag and it stay mounted.
4. If special mounting (such as riveting or welding) is required, the integrator does this himself. These tasks look easy and safe, because the licensed welders who do it make it look easy and safe.

5. The integrator assumes that throwing loose tags on an asset and running it through a reader is a good test. Orientation matters; what works on top of a box might not work on the side.

6. The integrator recommends a hardware device (handheld, static, tag, etc.) without field testing it first. The device may not even work with your configuration. Make sure you have the best reader(s) for the job. (Note: If the integrator must create strange work-arounds in order to get the data where you need it, it is not the better reader.)

7. The integrator does not guarantee that precoded tags (especially in the thousands) are uniquely encoded, and that the integrator will absorb any costs involved in correcting the problem.

8. The system loses data. Ensure that any data gathered for testing and certainly for the initial go-live is backed up.

9. The integrator convinces the users that RFID is magic and can solve all their asset-tracking needs, before conducting an in-depth analysis. The truth is, sometimes bar codes make more sense than RFID.

10. The integrator (or customer) goes into a project without rudimentary understanding of the technology. An experienced integrator will know that UHF does not work well on metal without special configurations or form factors, and that HF only has about a 6-inch read range.

11. The integrator believes that a spectrum analysis is unnecessary "because very few things interfere with this frequency," or some other excuse. Even if the integrator is experienced in a certain environment (e.g., a data center), there are unknown factors in each physical location that can interfere with various frequencies. Be certain the analysis covers at least a 24-hour period in or near the area where RFID is going to be used.

12. The integrator sends a U.S.-manufactured reader (or tag) overseas, without checking the standards of that country. Some countries have stricter standards on frequency, power setting, RoHS compliance, and so on. For example, in some countries, the 900 MHz frequency spectrum is completely illegal to use.

13. The integrator finds an overseas partner without verifying the RFID expertise, which partner proceeds to commit blunders #1–12.

14. The integrator installs the RFID equipment without prior experience. "Sure we know how to put the system together. We know how to put

the tags on right, configure the readers, and position the antennas." But some of us just aren't that handy with tools. Whenever possible, an integrator should at least try to work with clients to get their current facilities' personnel or contracting company to assist with installation.

15. Ordering the wrong tags for the assets/environment. Nothing screams "remedial RFID" like someone who puts RFID label tags on file folders, puts those file folders in a metal cart to be pushed through a fixed portal and then wonders why the tags don't have reliable read rates.

---

## SIDEBAR 2: HOW TO SELECT A WIRELESS INTEGRATOR

The growing demand for Wireless infrastructures is creating a lengthy list of self-proclaimed experts and integrators. Alas, there is no accreditation/certification/licensure to verify their expertise.

This is not new; since Walmart and the DoD made their RFID mandate announcements, the number of companies that offer RFID services grew exponentially. There were a lot of fish in those waters; as of 2004 when the announcements were made, there were more than 10,000 Walmart suppliers and 40,000 DoD suppliers. Combined, the two enterprises represented six percent of the U.S. GDP.

Of those suppliers, few understand the technology well enough to make informed purchasing decisions. As these suppliers struggle to become savvy shoppers, they become targets of the pseudo-service providers. Following are some recommendations we strongly suggest when selecting a Wireless service provider:

1. Ask how many full-time, experienced Wireless resources the company has. The key word here is "experienced." Web sites are easy to build; experience is hard won. Experienced integrators are very up-front with customers as to their areas of expertise and what services they can realistically provide.

2. Obtain résumés of the people who will be working on the RFID/RTLS project (where applicable). Ask if those individuals are employees or

contractors. Ensure that your services contract specifically states what role the experienced resources will play. Be certain the integrator will not just be assisting by telephone.

3. Ask for a presentation that will provide an overview of the Wireless solutions the integrator has developed for other customers. To be fair, keep in mind there will be a nondisclosure agreement in place with the other customer so details may be left out. A legitimate company should be able to state what equipment was used, how it was integrated, what challenges were overcome, and the current status.

4. Watch for "smoke and mirrors" demonstrations. Louis once toured a facility with RFID readers installed in dock doors. He took some tags out of his briefcase, waved them next to the antennas and watched the screen, and absolutely nothing happened. The readers were connected to a network switch, but the network switch was not connected to the computer.

5. Don't just ask for customer references; verify them, thoroughly.

6. Finally, ask with which hardware manufacturers the integrator has relationships. If the hardware provider certifies its partners (this is common in RFID), verify that the integrator is certified.

---

## SIDEBAR 3: WIP MANAGEMENT, RFID, AND ROI

Still unconvinced that RFID can bring ROI?

Consider this: work-in-process manufacturing environments are utilizing RFID technology and despite the "in-process" context, there is nothing unfinished about these results. CFOs across the manufacturing industry are waking up to the fact that RFID can dramatically streamline automated processes to increase quality and throughput.

WIP projects are common in the automotive industry. If you have never set foot in an auto plant, let us paint the picture. The assembly process is divided into multiple stages. In stage one, the vehicle chassis is placed onto a "carrier"; this is a steel frame with wheels. (Remember, the chassis has no wheels at this point but needs to move around somehow.)

As the vehicle-in-progress moves from stage to stage, additional components, including radios, windshields, and seat belts, are added onto the chassis until it is a finished car. Depending on the car being built, there may be hundreds of stages. Not all auto manufacturers assemble this way, but you get the idea.

## RFID for Work-in-Process

Here is where it gets interesting.

A typical car plant may have thousands of partially assembled vehicles throughout the facility and it is extremely difficult to differentiate them without reading the serial number. The same applies for the components going on the vehicle.

Years ago, manufacturers began using bar codes to improve identification and tracking. But bar-code printed labels simply do not hold up well enough under manufacturing conditions, and bar codes etched in metal do not read accurately. Also, as components are added, the line of sight to the bar code can become obstructed, therefore making it nearly impossible to obtain an already difficult bar-code scan.

In a typical WIP application, an assembly line worker using a bar-code gun scans the vehicle bar code to let the manufacturing system know that it has entered a stage. This manual process may take only a few seconds (assuming an easy scan), but those seconds add up to hours over the course of a week, and possibly weeks over the course of a year.

The real glitch comes when someone forgets to scan the bar code. Not only can this omission cause mistakes in tracking and metrics, it can also mess up a special order. Most manufacturers build special-order vehicles on the same line as other vehicles. If an operator at stage 17 makes a mistake, it may not be realized until stage 27. At that point, the vehicle must be removed from the line altogether and taken to "quality hold," where it can be diagnosed, repaired, and placed back into the appropriate stage to complete the assembly. An error such as this can add thousands to the cost of a single vehicle and eliminate any profit altogether.

Further tracking challenges arise if the vehicle is removed from the line due to an exception, such as damage that requires repair. The vehicle bar code must be scanned again once it is moved to the repair area. This rescanning is one of the most commonly forgotten actions.

## Enter RFID

RFID technology replaces, or complements, the bar-code technology because it offers benefits such as:

- Greater durability of the RFID tag than a bar-code label
- No human intervention for scanning
- No requirement for line of sight capability
- Faster and more accurate read rates

Manufacturing environments are harsh places for equipment. In these environments, RFID tags must be encased in plastic or another durable material to help them withstand dirt, oil, washing, temperature variances, and physical impact. The tags have screw holes for mounting directly onto a metal carrier.

## RFID Metal Mount Tags for Work-in-Process

Purists looking for bar-code-like utility can mount UHF RFID tags directly on metal, as long as there is a minimum 1/8-inch distance between the antenna and the metal. Otherwise, the tag will short out and not function.

Louis implemented RFID at an auto manufacturer, at a cost of about $2.50 per tag. At first blush that seems costly, but the tag, like the carrier, is used hundreds of times. So the per-use cost is less than a penny.

RFID antennas are mounted at each stage. At stage one, the bar code on the vehicle chassis is scanned and is associated with the number on the carrier's RFID tag. After that, there is no longer a need to scan the bar code. The RFID reader automatically notifies the manufacturing system when a vehicle enters or exits a stage. For special orders, the RFID reader is connected to a computer that displays exactly what tasks are to be performed for the specific vehicle as it enters the stage. Once a vehicle leaves a stage, the stage number is recorded so that if a vehicle is taken off the line, the system knows exactly to which stage to return it.

What gives these RFID solutions an ROI? First, there are measurable benefits from time savings and error reductions. Secondly, the project costs are known. They are fixed assets (readers, tags, cables, etc.), installation and integration services, and minimal ongoing support costs. Based on these savings, it does not take a CFO to determine how long it will take before the ROI occurs. An auto maker can achieve 100 percent ROI in as little as 12 months, because errors involving vehicles can cost thousands of dollars each.

An unplanned benefit that RFID implementers typically observe is the psychological benefit to the operators. This is not some touchy-feely nonsense; bear in mind one of the goals of Lean is worker satisfaction. Production operators are grateful to have the burden of bar-code scanning lifted, because they need not have to worry about a mis-scan that results in mistakes for which they are accountable. Not only are the operators more efficient, they are happier because their performance can only improve. It would be hard to put a dollar figure on that return on investment, but it is one that auto makers think is valuable indeed.

## ENDNOTES

1. Strauss, J. et. al. 2008. *E-Marketing*, 5th edition. Upper Saddle River, NJ: Prentice Hall.
2. Sharlach, A. and B. Grosswald. 1997. *The Family and Medical Leave Act of 1993*. Chicago, IL: University of Chicago Press.
3. Liard, M. and S. Schatt. 2008. *Annual RFID End User Survey*. ABI Research, Inc.
4. Liard, M. and S. Schatt. 2008. *Annual RFID End User Survey*. ABI Research, Inc.
5. Liard, M. and S. Schatt. 2008. *Annual RFID End User Survey*. ABI Research, Inc.
6. "Total cost of ownership for various computing models." © 2007, Principled Technologies, Inc. White paper.

# Index

# The Authors

**Dann Anthony Maurno**'s 25-year career in business and technology journalism began with a bachelor's degree in chemistry, progressed into engineering positions at Raytheon and Genetics Institute, took a turn into publishing and marketing with Factory Mutual and Lilly Software, and culminated in analysis for such groups as Decisions Resources, the Economist Intelligence Unit, and International Data Group (IDG). He continues to pursue field science as an amateur paleontologist, and his field experience includes digs at Olduvai Gorge in Tanzania at sites developed by Dr. Louis Leakey. He lives in Salem, Massachusetts.

**Louis Sirico** is the founder of The RFID Network, a global community dedicated to RFID and related technologies with a distribution of over 100,000 professionals. He is most widely recognized as the host and producer of *The RFID Network* video series and has authored over 200 published articles. During his 25-year career, he has developed RFID solutions for Fortune 100 companies both in North America and in Europe as well as the Department of Defense, and the Department of Homeland Security. He is credited with hundreds of professional speaking engagements as well as television and radio appearances and provides industry analysis and RFID subject matter expertise to CNN, CNBC/MSN, CBS News, *The New York Times*, Fox News, the MIT Enterprise Forum (mitef.org), *SAP Information Magazine*, *Inbound Logistics Magazine*, and many others. Louis can be found on the Web at http://RFID.net.